The
QRP Scrapbook

The very best projects from the G-QRP Club magazine SPRAT 2012 – 2017

Edited and compiled by
Steve Telenius-Lowe, PJ4DX

THE JOURNAL OF THE G QRP CLUB

DEVOTED TO LOW POWER COMMUNICATION

Radio Society of Great Britain

Published by the Radio Society of Great Britain, 3 Abbey Court, Fraser Road, Priory Business Park, Bedford MK44 3WH. Tel: 01234 832700. Web: www.rsgb.org

The G-QRP Club may be contacted at the following address: Steve Hartley, G0FUW, Chairman G-QRP Club, 5 Sydenham Buildings, Bath BA2 3BS.

Published 2019.

© Radio Society of Great Britain, 2019. All rights reserved. No part of this publication may be reproduced, stored in a retrieval system, or transmitted, in any form or by any means, electronic, mechanical, photocopying, recording or otherwise, without the prior written permission of the Radio Society of Great Britain.

ISBN: 9781 9101 9379 2

Compiled by: Steve Telenius-Lowe, PJ4DX
Cover design: Kevin Williams, M6CYB
Production: Mark Allgar, M1MPA

Printed in Great Britain by CPI Antony Rowe of Chippenham, Wiltshire

Publisher's Note:
The opinions expressed in this book are those of the author(s) and are not necessarily those of the Radio Society of Great Britain. Whilst the information presented is believed to be correct, the publishers and their agents cannot accept responsibility for consequences arising from any inaccuracies or omissions. Please check the G-QRP Club website at http://www.gqrp.com/sprat.htm for any corrections to articles published in *SPRAT*.

Contents

Preface, by Steve Hartley, G0FUW 4

1. Transmitters 5
2. Transceivers 33
3. Receivers 51
4. Antenna Systems 103
5. Test Equipment 133
6. Miscellaneous 154

Index of articles 237

This book is dedicated to the memory of
Rev George Dobbs, G3RJV, the founder of the G-QRP Club

Preface

This second volume of the 'best of' *SPRAT* is a testament to two things: firstly the readiness of G-QRP Club members to share their ideas and achievements and, secondly, to our inspirational founder, the Reverend George Dobbs, G3RJV, who passed away in 2019.

All of the circuits and project ideas contained here were freely given to *SPRAT* by their authors and many go on to provide support to those who reproduce and / or develop them further. The spirit of sharing and supporting is as strong in the G-QRP Club today as it was when it began over 40 years ago.

George fostered that spirit through his own selfless sharing and infectious enthusiasm. Amateur radio without George is an unknown experience for me – he has always been there; George was the catalyst for me becoming a radio amateur. One wet bank holiday weekend I bought a copy of *Practical Wireless* and in there was an article that showed how you could build a QRP transceiver at a reasonable price. The author (guess who) made it sound so understandable, so achievable that it left me thinking I can do that – and I did. I joined the G-QRP Club before I was licensed and thirty-something years later, I find myself as its Chairman.

The G-QRP Club and *SPRAT* are George's legacy and the current club committee is determined to continue the work in a way that George would have approved of. He will be missed but his legacy will continue and we will all do well to follow his fine example.

I hope you enjoy this collection and, if you are not already a member, that you will join us to enjoy more *SPRATs* and join in the sharing community that is the G-QRP Club.

Steve Hartley, G0FUW
Bath, August 2019

Steve Hartley, G0FUW (left) with founder of the G-QRP Club, the late Rev George Dobbs, G3RJV, Spring 2015.

1 Transmitters

40m CW Transmitter
Barry Zaruki M0DGQ, 26 Heathfield Rd, BIRMINGHAM. B14 7DB

The G-QRP Club sells a good range of 40M CW xtals at a very reasonable price. Here is a very cheap and easy to build transmitter using these xtals. Power output is approximately
6 Watts for a supply voltage of 13.8 VDC (if you experiment with the Pi output values and FET bias then over 10 Watts can be had, but a large heat sink will be required for the FET Construction is made easy by using a pad cutting tool for the circuit board - this tool is also available from the G-QRP Club. Easy to obtain components are used throughout the design.

The circuit consists of a colpitts xtal oscillator TR1 followed by a diode switch D1, D2, a driver TR2 and PA stage TR3. A zener diode ZD1 stabilises the DC supply to TR1 stage, this oscillator runs continuously to avoid any chirp etc. The oscillator output is keyed by a diode switch consisting of D1, D2 and their associated components. Capacitor C1 provides some envelope shaping of the keyed waveform; it is isolated from RF by L1. On key down C1 gradually charges up towards the supply rail via D4 and R1, thus D1 and D2 become forward biased allowing the oscillator signal to pass to TR2 without a steep rising edge of the waveform. On key up, C1 discharged via R1, R2 the time constant of which eliminates a sharp falling edge of the waveform. Tr2 is biased to class B via D3. Its collector load consists of L2, VC1, a parallel resonant tuned circuit. A low impedance tap on L2 feeds TR3 gate. The gate is heavily swamped by low impedance bias network and L2 tap, this is to overcome the 150pF or so of gate capacitance that would otherwise be difficult to drive. TR3 is biased for a standing current of 200 - 250mA by VR1. The drain load for TR3 is a Pi network, the 80pF (approx) of output capacitance exhibited by TR3 is taken into account for the network component values. A low pass filter (L5 and L6) follows the Pi network.

A two pole two way switch is used for antenna switching, DC supply to the transmitter and also a mute voltage for a RX IF. A matching RX is currently under construction for use with this TX. As mentioned earlier, it is possible to obtain 10 Watts output simply by changing the 880pF capacitor in the Pi network to a 1200pF and increasing TR3 standing current to 800mA. However, I do not recommend this unless you fit a large heat sink to TR3 as it does run very hot at this power level. The receiving
station is unlikely to hear any difference between 6 and 10 Watts Tx power, I run mine at 6 Watts for a total current drain of 750 mA, if you require 5 Watts exactly, reduce Tr3 bias voltage accordingly.

Alignment of the finished transmitter is simple, with no drive set the standing current of TR3 to 200 - 250 mA. Apply drive and adjust VC1 for maximum output power. Adjust the xtal trimmers for exact xtal frequencies, these trimmers were only included in case several of these transmitters are built for use in a net and all produce the same TX frequency. You may note an extra transistor on the circuit board, this is used as contact closure to ground

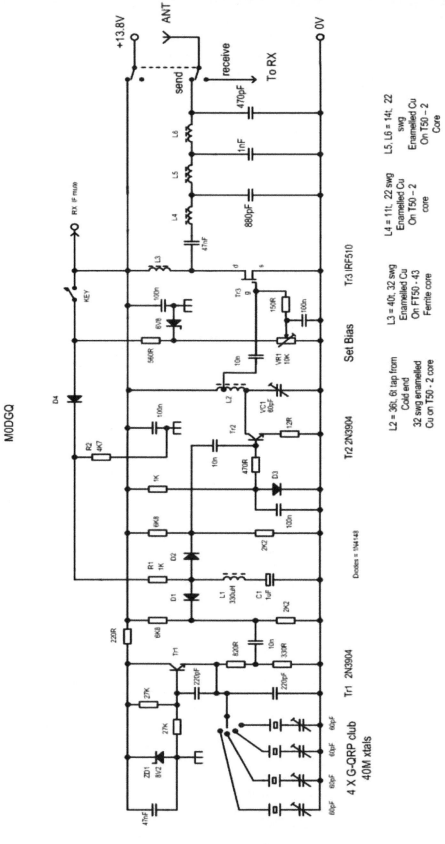

PART 1 TRANSMITTERS

key switch so my electronic keyer can be used with this set. The front and back of the case is made from some copper clad board, the circuit board forms the bottom of the case. The lid of the case is made from a perforated aluminium ceiling tile.

QRP SCRAPBOOK

LATATUN – Spanish Tuna Tin Transmitter
Lluis Terres Salto EA3WX, Apartado de Correos 149, S-25080 Lleida, Spain

One of the great masters of QRP, Doug DeMaw, W1FB, designed and published in QST magazine this transmitter assembled inside a small can of tuna and called it TUNA TIN and has become widely known within the world community of QRP. That was as far as 1976 and since then there has been multiple versions and modifications.
In the wake of the TUNA TIN American, we wanted to make a Spanish version of similar characteristics, but going to the assembly with SMD components. The circuit is not original at all but was set up initially by Rev. George Dobbs, G3RJV in 1998 (see SPRAT 96)

DESCRIPTION OF THE CIRCUIT
The circuit consists of a crystal oscillator with the transistor T1. The key is responsible for giving ground to the transistor so that the oscillator starts to operate.

In order to have a slightly variation of frequency, the crystal is mounted in series with a coil L1, 3.3 mH followed by a variable capacitor of 50 pF. If you do not wish any change can remove these components, bypassing the corresponding ground pin of the crystal quartz.
The coupling to the final step is performed by means of L2-L3.
The final step consists of T2 and the circuitry attached. T2 is PZT2222A type that it is still the classic 2N2222 in SMD format but with a power dissipation of 1 W.

It can be seen that the collector of this transistor is soldered to a wide copper surface; this is so in order to have a cooler surface to avoid the classic cooler bigger.

At the far end of the PI filter, which is not connected to the antenna, we have the capacitors C10, C11 and L7 coil forming a resonant circuit in series to carry the signal from the antenna to the receiver input, being tuned to operating band. The diodes D1 to D4 cut the excess signal that may be introduced to the receiver during transmission. However, it is still good that the receiver will get something of the transmitted signal as it will help us monitor our transmission as if we had a side tone oscillator.

ASSEMBLY

Although the assembly of SMD components may seem difficult and complicated, actually is not quite true. You just have to have patience, a bit of order to avoid mixing components (and not lose them!), a good magnifying glass and a fine tip soldering iron.

Another advantage of SMD components is its low price that can buy it in quantity without the costs are excessive.
The printed circuit board is attached to the can by means of a spacer and two screws that secure it to the bottom.

PART 1 TRANSMITTERS

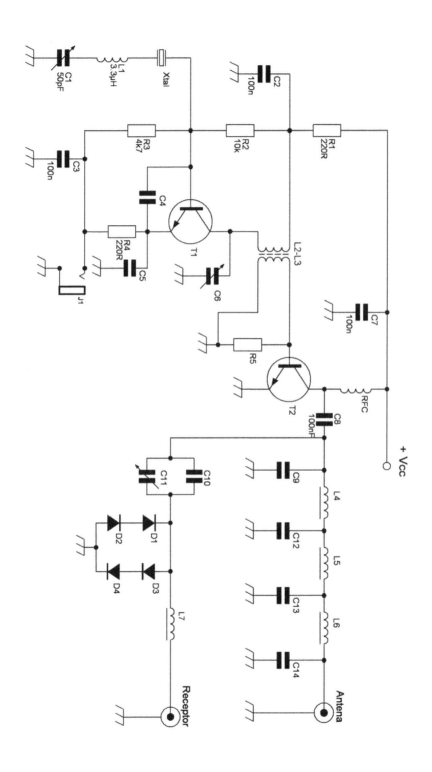

QRP SCRAPBOOK

TRANSMITTER TEST

Once the transmitter assembled, prepare the equipment for their operation and adjustment, i.e. load resistance of 50 ohms, power supply or battery, a wattmeter and the key.
If there is no short circuit between positive and negative of power supply line, connect power and control current with an ammeter.
If all these cautions are taken and all is OK, just push the key down and observe what happens. If we pass what I call SMOKE TEST, i.e., NO SMOKE has come from nowhere, we should adjust the transmitter.
One can say that the only major adjustment to do is tweak the C6 variable capacitor until the oscillator start-up; achieved this, just have to fine-tune both C6 and L2 to get maximum power output.

Connecting a receiver of the right band, in our case 14 MHz, to the output connector of reception, you must adjust the variable capacitor C11 to achieve the highest level of receiving signal. This capacitor may interact with the transmitter circuit, will have to compromise between the maximum received signal and the maximum output power.

In our case, and for a supply of 13 V, the output power is 700 mW. We note that with a supply of 12 V the power output is 500 mW.
It was possible to measure the signal level of the second harmonic (at 28 MHz) being 54 dB below the fundamental signal of 14 MHz

PART 1 TRANSMITTERS

EPILOGUE
With this setup I want to pay tribute to Doug DEMAW, W1FB who was the creator of this lineage of the TUNA TIN and died in 1997.

Doug began as a farmer in Michigan and managed to become an engineer, founder of several companies engaged in the manufacture of radio equipment for the aviation industry. He was a prolific writer of books and articles for the ARRL, which was also the manager.

Home made panel tip
David Smith G4COE, 54 Warrington Road, Leigh, Lancs.

Using thin coloured card of your choice, print the designed and laminate for this I use Proteus Labcentre pcb layout cad for this, there are plenty free electronic panel design programs on the web, there's even programs for making your own meter scales!

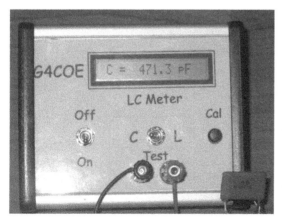

After printing trim the card to suit, then laminate the card and again trim the lamination to suit the card size.

For LCD display's and such, cut the card slot as per requirement then laminate so that you laminate over the cut out, leaving a protective window, this hides all the scratches and marks and gives the item a professional look far, far better than using stickers.

AUTUMN 2012 SPRAT 152

QRP SCRAPBOOK

The G4GDR 6CH6 Junk Box Special
Described by Coin Turner, G3VTT. (g3vtt@aol.com)

Adrian G4GDR has always been a prolific valve QRP operator and has obtained good results with his single EL91 receiver. I asked Adrian if he could supply details of his matching transmitter using a 6CH6 video valve.

"Hi Colin! This little rig was built from the junk box using the 6CH6 as it was to hand. This valve was designed as a video amplifier power pentode. I've had a lot of fun with this little TX and have worked G, GW, GM, EI, PA, ON, OZ, LA, F, DJ, OK, and IK6 with the last station being the best DX. I load it into an atu although on occasions I worked straight into my inverted G5RV.'

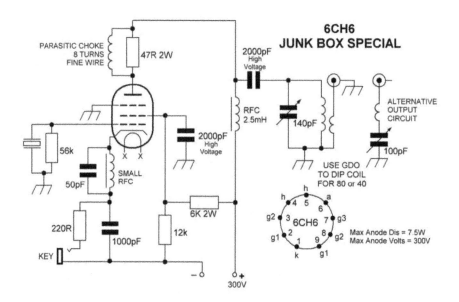

'I claim no originality for the circuit. Using my one valve receiver and this transmitter I've had hundreds of good QSO's on 80m. The transmit note is very good and always gets T9 reports. Reports into G are usually 559 and from Europe at night I get 569 or 579. On tune there's one dip point which corresponds with maximum power. I use a good quality 3560 crystal and get around 3 watts output." *What more do you need? The anode dissipation is around 7.5 watts for the 6CH6 and the maximum anode voltage is 300 volts. Some experimentation may be needed for the tank coil but you could try 22 turns for the tank coil on a 1.5 inch former with a three turn link over the cold end. An 18 turn link could be used as shown as an alternative to couple to a long wire and counterpoise arrangement. The stabilised screen volts and oscillator capacitor values ensure a clean pure output, just like a little whistle. Our thanks go to Adrian for his circuit and to George RJV for the diagram. G3VTT*

PART 1 TRANSMITTERS

40 METRE WSPR TRANSMITTER
Chris Osborn G3XIZ, g3xiz@yahoo.co.uk

This is a simple circuit comprising some 30 easily obtainable components.
It may be built in an evening or two and is simple to set up. The circuit was inspired by my ZL2BMI DSB transceiver (Ref 1) which required very little modification to make into this unit. Operating as a WSPR QRP beacon, it may be left unattended, bringing in reports from many parts of the world. With a little extra effort it could be made into a WSPR TRX.

WSPR
WSPR (Weak Signal Propagation Reporting) is a most useful and fascinating transmission system. Its use and principles have been described before in 'Sprat' (Ref 2) so a brief recap should suffice. WSPR is an automatic beacon-type mode, not a QSO mode and utilizes a PC to transmit and receive very weak signals. Transmissions last for periods of just under two minutes and the user selects the ratio (percentage) of transmit to receive time.

Signals are reported via the internet which identifies the reporting station, distance, azimuth and received signal strength. During the course of a single evening one may log hundreds of reports from dozens of stations. Needless to say, the system is most useful for transmitter and aerial tests and for gauging propagation conditions.

CIRCUIT DESCRIPTION
Although WSPR is used mainly with 'black-box' SSB transmitters this unit employs a simple double sideband (DSB) circuit.

The unwanted sideband should cause minimum QRM at the low power level used and the loss in radiated signal compared to an SSB signal (-3 dB) will be hardly noticeable under normal QSB conditions. The LM602 mixer IC generates the 'carrier' frequency via it's associated crystal oscillator. The audio input is derived from the PC soundcard and the level is adjusted and fed to pin 1. A double sideband signal is taken from pin 4, and amplified by the buffer Q1 and the PA transistor Q2. With the DSB signal fed to the aerial via a suitable ATU the power output was found to be about 500 mW pep

CONSTRUCTION
The entire circuit was mounted on a small piece of veroboard, attached to a ground plane of copper clad board using stand off pillars. Finally the assembly was encased in a small aluminium box having 2mm sockets for the supply, a PL259 socket for the aerial connection and a 3.5mm jack socket for the PC input. An LED indicator was mounted on the front and is of course optional.

QRP SCRAPBOOK

TESTING
Check the veroboard visually for any dry joints and short-circuited tracks, then connect the supply and turn on. The standing supply current should be about 45mA (35 mA if no LED is used)
Check that the circuit no-signal voltages correspond approximately to those shown. Ensure that the crystal oscillator is working and generating approximately 7038.75 kHz. If necessary adjust VC to pull the 7040 kHz crystal down to this frequency. Next connect a dummy load across the aerial socket and inject an audio signal, approx 1.5 kHz at 1V p-p into the jack socket. With an oscilloscope connected across the dummy load the displayed signal should be a close approximation to the classical double sideband waveform.

SOFTWARE
Review the WSPR website (Ref 3) and download the WSPR software. It is quite intuitive although I found it beneficial to experiment with receiving WSPR signals before attempting transmissions, noting that the 40m WSPR sub-band is between 7040.000 and 7040.200 kHz

ON AIR TESTING
Connect the transmitter to the aerial via the ATU and the front input jack to the PC sound card line output. Open the WSPR transmit/receive software and set the band for 40m. Now insert the details required in the 'Setup' drop-down box, including callsign, location and power which should be 23 dBm (0.2 watts)

In the 'Frequencies (MHz)' 'Dial' dialogue box on the WSPR transmit window type in your measured crystal oscillator frequency e.g. 7.038750 (or a good guess if a counter is unavailable)
Choose a transmit frequency in the WSPR sub-band, ideally clear of other stations and type this into the 'TX' dialogue box e.g. 7.040140

Transmissions occur in 2 minute sessions at the start of even minutes so it is important for the PC time to be accurately set. Although there are various methods of doing this, the internet clock, accessible from the 'Toolbar Time' in the 'Windows' system is probably the most convenient.

The facility to set the percentage of time that you will transmit, can be initially set to 100% to avoid a long wait while testing, then afterwards reducing it to a more sociable 25%

Once the software has initialized the bottom right hand dialogue box should change from 'Receiving' to 'Transmitting . .(with callsign information)'
At this point adjust your ATU for minimum SWR.
Finally revue the WSPR website for reports on your signal.

PART 1 TRANSMITTERS

RESULTS
In the first 24 hours, transmitting with about a 25% duty cycle I received over 800 reports from some 30 stations. Conditions were not too good on 40m but my best report was from LA9JO at over 2000 km. On the second day I had a DX 'spot' from VK7BO at 17288 km so world coverage would seem possible.

I keep a log of these reports in an 'Excel' spreadsheet as a useful reference against any aerial and transmitter changes. There is no reason why this circuit should not be used on other bands provided suitable crystals may be pulled to the appropriate frequencies.

As with so much that is QRP: experimentation is the key to self-tuition and also gets the results.

References

1. 'Sprat' Spring 2011 "ZL2BMI extra simple DSB 80m Transceiver (ZL2BMI)
2. 'Sprat' Autumn 2009 "WSPR- what is it and what it can do" (G3XBM)
3. www.physics.princeton.edu/pulsar/K1JT/wspr.html

QRP SCRAPBOOK

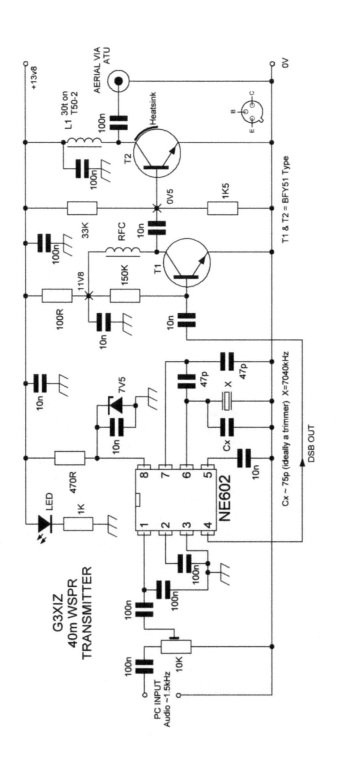

G3XIZ 40m WSPR TRANSMITTER

PART 1 TRANSMITTERS

A 6V6 CO/PA TRANSMITTER
Derek Thom G3NKS, Cheltenham, Gloucestershire.
g3nks@blueyonder.co.uk

The inspiration for this project came from the Summer 2011 issue of "Sprat". An article by Colin G3VTT described a two valve "straight" receiver; the two valves both being 6V6 beam tetrodes. As Colin said in the article, some might think this valve is overkill for such a receiver but apparently it makes a good regenerative detector with low distortion.

I thought of building such a receiver, but decided first to build a companion transmitter; companion in the sense that it too would use two 6V6s and be built into a die-cast box. In the autumn of 2011 I started to research possible designs but found nothing suitable. So I based my design on a couple of circuits from early RSGB publications which I adapted for 6V6s. Chassis bashing commenced in October but the work came to halt when the weather turned cold: my workshop cum second-shack is a shed in the garden. In mid-February work resumed (cold or no cold!) in order to complete the project in time for the Cheltenham club's annual Constructors' Exhibition in March. CO/PA stands for Crystal-Oscillator/Power-Amplifier; a typical arrangement for an amateur transmitter in the 1930s and 40s. Needless to say, it's a CW only rig.

After some experimentation the final circuit is as shown in the accompanying diagram. Like Colin, I've wired the heaters in series to cater for /P operation from a 12V battery. The RF by-pass capacitors are all disc ceramic and could be any value in the order of 10nF, 300V working. All the resistors are half-watt except those in the screen grid circuits which are 1W. The 50k pot for adjusting the power output level is also rated 1W. All the RFCs are 2.5mH. The pi-network coil, L1 (20uH), consists of 32 turns of 18 swg wire space wound on a 1½ inch ceramic former and tapped 13 turns from the valve end for 40m. The "Load" capacitor should ideally be a double gang 2 x 500pF variable, but I didn't have a suitable one to hand at the time, hence the 500pF fixed capacitor in parallel with the 480pF variable. A power supply, which I built some years ago, provides about 250V dc for the HT line. If you are contemplating building such a rig please take care with the high voltages involved, they are dangerous!

Luckily I've several old style crystals (eg BC-610/FT-171 and 10X types) which are suitable for a valve oscillator. The trimmer capacitor in the grid circuit of the oscillator was adjusted to give reliable oscillation but the setting is not critical. The rig produces a maximum of 5W of RF. From my second-shack/workshop I've made QSOs around the UK on 80m and into continental Europe on 40m using 75ft of wire at fence-top height and an old Drake 2C receiver.

My aim now is to build the 6V6 receiver featured in Sprat plus a matching power unit and ATU, all in separate die-cast boxes. Come to the 2013 Cheltenham club's Constructors' Exhibition to see if I manage to fulfil this ambition!

QRP SCRAPBOOK

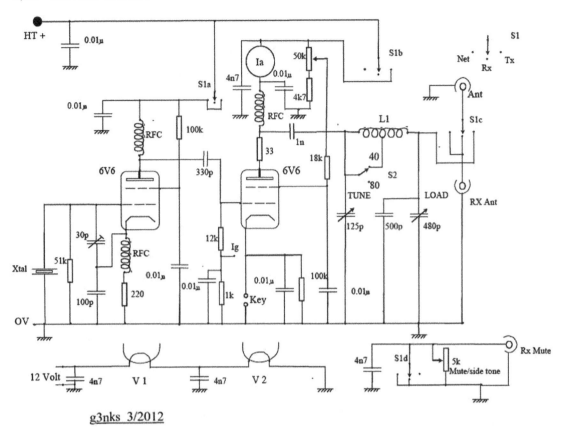

g3nks 3/2012

UNDERSIDE VIEW

PART 1 TRANSMITTERS

Modular Transmitter
Peter Howard G4UMB 63 West Bradford Rd Waddington Lancs

This simple 80m transmitter was made to try out all the crystals I had pulled a few kHz. to other frequencies which I covered in my last article. I decided for a change to build using Plugin Boards . The plugs and sockets are cut pieces from turned pin IC sockets

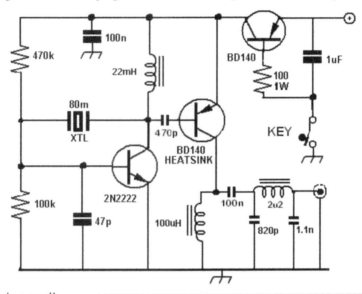

The base is a small tobacco tin. It may be practical to make the transmitter work on another band by changing the crystal and PI filter plugin board? The output power can be adjusted by altering the supply voltage. If you run it on 19v at 5W as I do sometimes using a 2W 100 ohm resistor in series with the key would be better as my quarter watt one gets too hot. Also note the BD140 needs a good heat-sink.

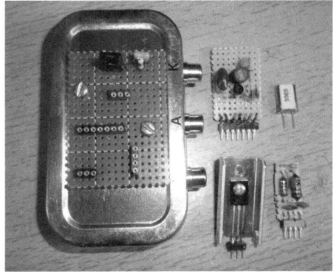

QRP SCRAPBOOK

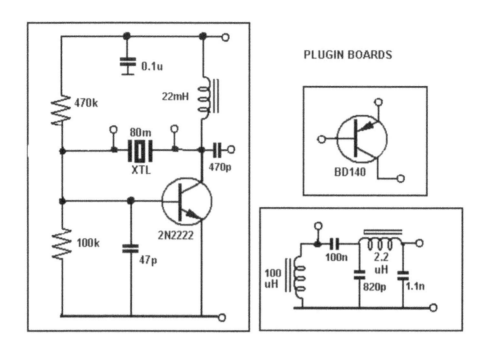

PART 1 TRANSMITTERS

The KR80 Short Aerial 20 Metre Band QRP Transmitter
Keith Ranger G0KJK

Not few G QRP C members find themselves living today in circumstances where efficient outdoor aerials cannot be erected, perhaps because of a very small garden or specific local restrictions. So does this mean it is virtually impossible to get on the air with QRP and enjoy plenty of good QSOs? Not with the rig and aerial to be described, a simple home-brew three transistor circuit that puts out 1.8 watts on 20 metres CW for 12 volts and 250 milliamps power consumption with the key down, easily matched to an indoor aerial of only 10ft draped around suitable objects on the shack wall (in my case, two hanging pictures, see diagram) using the easily constructed inbuilt aerial tuning unit (ATU). Does it sound too good to be true? Try it! And let me know your results! See below the call-signs of, and reports received from, the last twenty entities contacted with this almost non-aerial set up (none of these QSOs were rubber-stamp exchanges, they were all information-sharing one hundred percent contacts):-

DF6EW 539, K3SEW 569, F3BNX 559, IK2LFF 599, OK1XZ 599, EU6AF 559, E74X 599, YO3FRI 339, 9A9C 599, UA1AML 569, ZB3MED 599, SP0VD 599, SM7ZKI 579, YU1AKV 599, OM2VL 599, OE7PET 549, EA5DNO 579, EA6UN 599, LY2PX 589, SP4LNV 559.

Notice that no less than nine entities gave the little rig 599 and only one a report of less than R5 (and even this QSO~ was successfully completed). This shows that if you build this kind of a rig and use a well-matched short aerial you should get in reasonable conditions QSOs all around Europe, the Mediterranean islands and even on occasion into further afield places like North America. So we need not collapse in despair if we cannot erect a G5RV of Yagi antenna! Minimalist stuff does the trick and as out Club slogan puts it – "It is vain to do with more what can be done with less"!

The circuit of this little 1.8w power-house (called the KR80 after my initials and current age!) is hereby given. The VXO stage uses a series of T 50 2 inductors connected to a rotary switch. You may need to experiment with the total number of turns but this should give you full coverage of the 14060-14000KHZ CW section of the 20 metre band. Contrary to expectation, this rotary switch-selected set of inductors can provide stable results on all frequencies and with 1.8w still going into the aerial even near the band edge. I work many stations around 14010KHZ with this configuration. The ATU coils on the T 68 2 toroid are very straight forward in their winding. (please see diagram), the one turn on the binocular or pig-nose two hole ferrite balun core secondary should be tightly twisted as in the diagram. If this kind of a core is not available, the easily acquired FT50-43 ferrite ring can be substituted, with a twelve turn primary and a three turn secondary, for roughly comparable performance (and greater ease of winding). Again, please see diagram.

QRP SCRAPBOOK

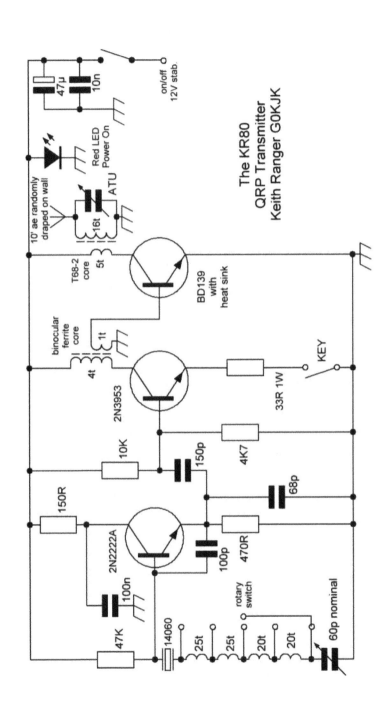

PART 1 TRANSMITTERS

Extra Notes

Further to the article itself I would like to add this for your information:-

1. The wide-range VXO is based on an article I once read in SPRAT (I cannot recall which issue) on how a series of small inductors can produce stable wide-ranging frequency change when a large inductor of the same total value will not so work (I think the article was about 80m). This works very well in the KR80 (its name is based on my initials and current age!) circuit. The only caveat is that with all the inductors brought into play by the rotary switch the crystal might not fire unless the variable capacitor vanes are fully disengaged. After that the vanes can be fully engaged and stable results expected right to the band edge in terms of power output and T9 note. I work many stations in the lowest 10KHZ of the band.

2. The inbuilt ATU, configured as shown, is undoubtedly the secret of the consistently good reports I receive, very rarely less than 559. If a ferrite bead wound with about ten turns and a 0.1 capacitor takes the RF out of the BD139 collector instead, with or without a LPF, output power and matching are much poorer and on air results also not nearly as good. I worked all ten call areas of Japan from Hong Kong (some of them 2000 miles away) when I was with the church as VS6US and with only 1w of output power. What once worried me was absence of further filtering, but both there and here I have nbot had a single complaint about TVI of BCI. I have recently noticed that the G3IGU transceiver on page 202 of The Low Power SPRAT Book uses a very similar aerial coupling circuit. I am lost for an explanation of how and why this works so well with short indoor randomly orientated antennas but it sure does (does it perhaps function like a base-loaded whip?! Yesterday evening I reduced the aerial length to only 8 ft and totally changed aerial orientation around my two picture frames. Here is my resulting log for 8ft (only one of the a rubber stamp contact and in one case with a follow-on QSO giving that station a poorer RST than mine):-

HG30CW 1950hrs 599, E74X 2005 599, YT4T 2115 579, HG5DX 2128 599. I had a narrow miss with A45HR in Oman!

A hot tip for finding cheap variable capacitors – hunt around at the rallies! I find plenty there, more than I would ever expect to need to use!

If you hit any problems in putting the rig/ATU/aerial together, please feel free to contact me at keithcath@ranger144.fsnet/co.uk I am more than happy to explain points I have made as clear as mud!

QRP SCRAPBOOK

THE KR80 TRANSMITTER – ERRATA and ADDENDA
Keith Ranger G0KJK
<keithcath@ranger144.fsnet/co.uk>

As a result of enquiries received since the publication of the circuit and accompanying text describing this 20 metre QRP transmitter, I would like to clarify several points:-

1. The second transistor in the circuit diagram is not a 2N3953, it is as several members have correctly deduced a 2N3053. Because the emitter resistor is only 33 ohms, it should be given a crown heat-sink if you operate the transmitter from a higher power source than 12 volts. It works well with 15 or 18 volts and of course the RF power output is thereby enhanced. If instability is encountered at any power input level, fit a small ferrite bead on the BD139 base lead-out and a 100 ohm half watt resistor between the BD139 base and ground. RF power output should remain unaffected.

2. No value is given in the SPRAT no. 164 circuit diagram of the KR80 for the variable cap in the ATU. This can be any value up to, say, 100pf with the vanes fully meshed. The resonance point for 20 metres is quite sharp and will probably be encountered by you with the vanes only slightly enmeshed.

3. The circuit as printed should have had an 0.1 capacitor across the key for shaping and a further 0.1 capacitor from the emitter of the 2N3053 transistor to ground will slightly enhance inter-stage power transfer, up to 1.9 watts out may then become possible (270ma with the key down). However, the transmitter may yield from 1.3 to 1.9 watts depending especially on the particular BD139 device used.

4. I have been asked, can the KR80 be used on other bands? The answer is Yes, but only 20m does well with a short indoor aerial. I am finding that the length of choice for 20m is actually a precise 8ft3ins, which represents an eighth of a wavelength on 20, performing slightly better than either 10 or 11ft. 17 and 30m require an end-fed outdoor aerial of around 33ft and you'll need to put out some 45-66ft for reasonable results on 40 and 60. Also, add some extra inductors to your VXO rotary switch for better LF band coverage. The little 10 or 22 Micro-Henry axial chokes sold by Graham G3MFJ at Club Sales would be ideal.

Finally – persist! Conditions on 20 have not been good recently but I have still had some good QSOs, albeit with slightly poorer reports at times.

Good hunting to you all

PART 1 TRANSMITTERS

Three "Moxo" (modified OXO) QRP Transmitters
by Revd. Keith Ranger G0KJK   G-QRP-C #8040

When Keith first submitted this useful article he offered an overview of his experiments and the development of the "Moxo" transmitter. This was followed up with a set of circuit diagrams and attendant practical notes. Rather than make life more complex, I have kept the material as it was submitted – enjoy Keith's circuits!

In the Spring of 1987, licensed as VS6US as I worked with my XYL Catherine with the church in Hong Kong, I decided to try my hand at building a small and simple transistorised QRP transmitter, previous attempts having used valves. I came across a circuit I still think absolutely brilliant – the OXO transmitter by a man I'm sure we all hugely respect – George GM3OXX. I built it exactly as featured except for the PA transistor: a 2N4427 or 2N3866 device was not readily available so I used the locally easy to procure 2SC756, not a perfect equivalent but the best I could then find.

As in George's original article (please see Diagram 1) I took the output from the PA collector via a ferrite bead with some ten turns of wire and a .01 capacitor to a homebrew ATU and an end-fed wire some 30 feet in length suspended by a bamboo pole to resist Hong Kong's sometimes ferocious winds. I then called CQ countless times with no reply – and Hong Kong is a quite rare entity! Finally, Hiro JF4EPN in Hiroshima came back to a 14060khz CQ and gave me a RST329 report! The QSO was difficult but I still have his beautiful QSL card, featuring peaceful lakeside pagodas, so appropriate when one recalls how that city had to suffer the dreadful effects of the first ever atomic bomb.

After this, I looked again at G3OXX's output circuitry and decided to shift the ATU into the collector circuit as per Diagram 2. Then I wound it as pictured and tried this new configuration on 21mhz. My very first CQ drew an instant reply from Ryo JA6SDR in South Japan, some 1,200 miles distant, my report was 539 and an excellent, easily maintained QSO followed. I used the new ATU arrangement on 14mhz and had similar success, with a few more turns on the ATU coil but again with a 1:3 turns ratio. I called the new transmitter "The MOXO Mk 1" and worked all the call areas of Japan with it, from JA0 to JA9, the furthest some 2,000miles away and I still have all the QSL cards from them. Other entities were also worked. The ATU tuning is sharp, I never received complaints of RFI or TVI, and recommend its experimental use if the original OXO circuit (with which I am so impressed) lures you into heating up your soldering iron!

QRP SCRAPBOOK

Returning to the UK in 1989, and with the call G0KJK, I found things much harder. A W7 station who worked me in Hong Kong wrote – "My first VS6 in 58 years of hamming!", but now I was just a small signal G station whom folk did not fall over themselves to work. So it was back to the drawing board again and the result was MOXO Mk 2 (Diagram 3). The idea is so straightforward that I imagine others must have tried it, but sometimes we human beings do not see the obvious! Someone has said that "an inventor is someone who sees what everyone else has seen but thinks what no one else has thought" (I imagine our respected George G3RJV who is an expert on quotations must have used it himself somewhere!). So the idea came to me of using 2 PA transistors in parallel, as per the diagram. The ATU arrangement remains the same but with more turns on the coil at 1:3 ratio. Two 2N3053 transistors were the PA devices used, their FT of around 100mhz makes them ideal for 40 metres, the 2N1711 is a possibly even better choice. The new MOXO was connected to a 66ft wire and tuned sharply. I got QSO after QSO with it at around 2 watts output and the commonest report was 579.

My problem here in the UK has been that the only QSO outside Europe I have ever achieved with a MOXO Mk 1 or 2 is with an EA8 station in the Canary Islands who gave me 339 and QSL-ed me. So I decided to add a further stage to George G3OXX's original design (Diagram 4). This used a BD139 output transistor and with 12-18 volts input power easily generated 2-5 watts on all bands from 17-80 metres. It wasn't long before QSOs with Brazil, Ivory Coast and the USA were in my log. I still use a variant of this circuit today, my recently published in SPRAT's KR80.

Good luck if you try any of these OXO mods, please feel free to contact me if you have any queries, at keithcath@ranger144.fsnet.co.uk. Keep experimenting if initial success seems elusive, I did and it has reaped good dividends and given great pleasure. Good QRP hunting!

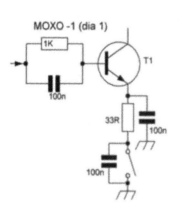

The original circuit of the OXO transmitter as featured in SPRAT. One change – I left out the keying transistor (BCY70) and placed the key in the PA emitter – as shown.
The transistor is the 2SC756, but there are many alternatives like 2N4427 or 2N3866. See note below on the 2N3866 from the club sales

PART 1 TRANSMITTERS

Diagram 2 is the Moxo Mk1 with the PA stage but no outboard ATU. 150pF polyvaricon variable,

1:3 ratio transformer. Try 3t/9t for 15, 4t/12t for 20, 10t/30t for 40, all on T68-2 core. Try 2N 4427, 2N3866 or 2N3053 OK for 20/80, But not so good for 15 or 17m.

PLEASE NOTE:
The club has a stock of 2N3866 transistors -
5 for two Second Class stamps
Or post free as part of a larger order

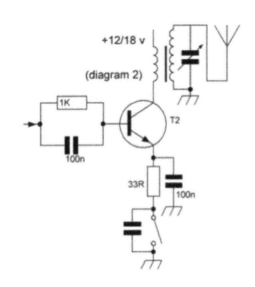
(diagram 2)

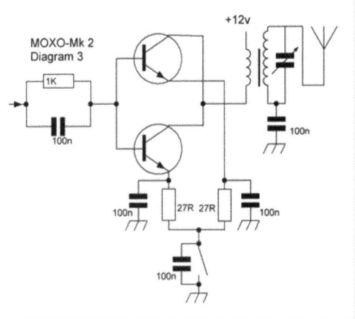
MOXO-Mk 2
Diagram 3

This is a very effective mod for 40m. Expect at least 2 watts out with 12 volts in and a 2N2222A as exciter.
More RF power is available with 15 or 18 volts.

Four 40m ATU
10t/39t on T68-2

QRP SCRAPBOOK

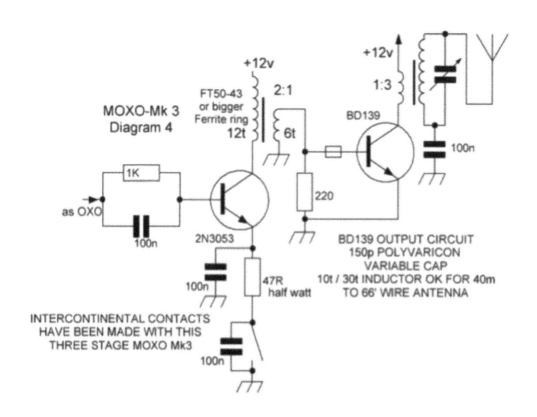

PART 1 TRANSMITTERS

Small Talk 160m Transmitter on FM
Peter Howard G4UMB 63 West Bradford Rd Waddington Lancs

For the past 10 years I've have a regular chat on 160M every week using 5Watts AM using my Small Talk Transmitter (See Sprat 133 for extra details) with a station about 10 miles away. Recently because of QRN I have had to use a loop ant. to receive and a long wire antenna for transmitting. However AM is still noisy although the signal to noise ratio improves a lot using the Loop. We found the answer to getting rid of QRN was to use FM instead. With adjustment of the squelch control on my Yaesu receiver I can now hear a good clean signal.

However the problem I had was I still wanted to run Homebrew but had no 160M FM transmitter so I set about modifying the Small Talk to make it suitable.

The circuit didn't need much work. It was a case of disconnecting the output of the LM380 mod. IC from the AM mixer IC to the gate of the VFO transistor. The two 6.8k resistors were chosen to allow enough audio through without stopping the oscillator. I had to supply the AM mixer IC with a new supply via the 330 ohm resistor. So a DPDT switch allows the transmitter to be used now on AM and FM. Unfortunately the only drawback is that the oscillator does not stay on the same frequency between switching because of the extra coupling on the VFO. So you can't switch modes during a QSO.

SMALL TALK (Sprat 133)
Modified for FM

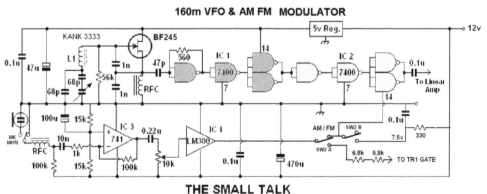

160m VFO & AM FM MODULATOR

THE SMALL TALK

Modified Limerick Sudden TX for 5262 kHz
John F Alder G4GMZ johnalder1@btinternet.com

The 60m band QRP CW calling frequency is 5262 kHz and a transmitter/receiver [TX/RX] need only operate within a couple of kHz of centre to cover most activity and indeed, the band [5258.5 – 5264.0 kHz in UK]. Rather simple designs of TX are practical and few more so than the Limerick Sudden [L-S] TX modules available from the GQRP Club although there is not one available for 60 m. The Club has made available the circuits and construction details on the webpage:
http://www.gqrp.com/Sudden_TX_Kit_manual_40m.pdf
and it is straightforward to plan design changes for experimentation and for members to purchase many required components also from Club Sales.

For this project the goal was an 8W CW TX matched to a small antenna centred on 5262 kHz to work initially with the base station Icom R9000 RX & LW antenna [35m wire at 4m above ground level and 25m counterpoise].

The centre frequency 5262 kHz is mid-way between 3560 and 7030 kHz and it is not certain which of the 80m or 40m TX modules would be best to modify. Study of the circuits quickly identified the frequency dependent components and almost arbitrarily it was decided to start with a 40m TX kit as it left me with a 7030 kHz crystal which will be useful in another project! To calculate the values of components required for 5262 kHz operation two approaches were used: a "half-way" estimate and "proper calculation"; in reality the half-way method proved quite adequate within experimental error. The chosen values that were different from the components supplied in the L-S 40m kit are shown in Table 1 against the part designation used in the component list for the kit.

The output Low-Pass filter values were calculated using the tables given in "Radio Communication Handbook" 10th Edn RSGB p.A4; and in "Building a Transceiver" E Skelton EI9GQ & E Richards G4LFM; RSGB 2014 p 46, for a 0.1 dB ripple 7-Pole Chebyshev LP filter which is the same component layout as used in the L-S 40m TX kit . The circuit was constructed on the kit circuit board using the toroids and wire provided and worked almost immediately. Component R6 was changed from 10R to 22R in order to bring drive adjustment into the middle-range of RV1 for optimisation. The frequency control of the crystal centred at 5262 kHz and tuned between 5261 and 5263 kHz after optimising with the trimmer on VC1. The handy power meter module supplied was beefed up a bit after initial measurements at 2W, using higher voltage-rating and powerdissipation components to cope with the anticipated 8W to be measured. An AVO-8 meter was used to measure output voltage of the module. Output power from the TX into the 50R dummy load was set and measured as between 1.3W with a 12.0V power supply and 2.5W with 13.8V; the transmit frequency did not alter significantly over this supply voltage change. Before connecting to an antenna the LPF was disconnected from the circuit and connected to the RF wattmeter module. The RF from a signal generator was input to the filter and monitored on an oscilloscope at the dummy load input. The filter parameters had been calculated for cut-off at about 6 MHz; the filter was seen to show peak transmission at 5.6 MHz; which by 6.4MHz was -20 dBV and by 7 MHz was -40 dBV on the peak signal. From 7 MHz up to 16 MHz the signals were too small to measure with the kit available. The completed TX terminated in 50R was set next to the RX and the harmonics of 5262 kHz up to 30 MHz were all well down in the noise from the RX; the filtering was considered adequate.

L-S TX: Changed Components for 5262 kHz Operation		
Component	Value Employed	Comment
R1	22 R	¼ W
C12	570 pF Foil	470+100
C13	1000 pF	Foil
C14	1000 pF	Foil
C15	645 pF	Foil 2x330
C18	150 pF	Silver/mica
L1	18 µH	Axial lead
L3	15 µH	Axial lead
L4	T37-6; 23 turn	30 SWG, 35cm
L5	T37-6; 22 turn	30 SWG, 35cm
L6	T37-6; 23 turn	30 SWG, 35cm
X1	5262.0 kHz	HC49/U

Use of the L-S 5262 kHz TX as driver for a PA
The power amplifier circuit chosen is by Drew Diamond VK3XU described in detail in Rad.Com. Handbook p 5.10 et seq, loc cit. It is simple to construct and is for a 1.8 to 10.1 MHz push-pull amplifier providing 5W using two IRF510 switching FET [available from Club Sales]. The only change needed was to optimise the components in the output filter for 5262 kHz operation and this was done with the aid of the table provided on p 5.10 loc cit. The coils for the LPF were wound on T68-2 toroids. The three inductors were wound with 18, 19 & 18 turns each; 40 cm each of 22 SWG enamelled wire leave 10-12mm tails. [Toroids and wire were from Club Sales.] The capacitors used were two each of nominal 600pF and 1000 pF foil type and the calculated filter cut-off frequency was between 6.3~6.6 MHz when the actual component measured values were put in, using the formula explained well by Skelton & Richards p 46 loc cit. The circuit was built into a metal sweetie box and optimised by driving it at 6 & 7 MHz from a 100 mW ex-MoD Test Generator.

The L-S 5262 kHz TX & PA were brought together and powered from the base station PSU. Output was to a Tokyo Hy-Power Labs Inc ATU and a Carolina Windom 40-10m antenna at 4m agl. Keying of the L-S TX was with a Redifon telegraph key which has a parallel contact that closes before the morse key contact is made. That was used to mute the R9000 thus providing full break-in keying. RF output at 5262 kHz was measured at ~7W into the ATU 50R dummy load from a 12 V supply and ~10W from a 13.8V supply. Total current from the PSU was ~80 mA quiescent and ~1A key down [12V]; ~2A key down [13.8V].
The Limerick Sudden 5262 kHz TX & PA have since proved to work well over many months with good reports from contacts into Europe and over UK. The construction was most enjoyable and this simple approach to home-brew 60m CW QRP operation can be recommended.

QRP SCRAPBOOK

SIMPLE 40m Transmitter
Peter Howard G4UMB, 63 West Bradford Rd. Waddington. BB7 3JD

A couple of years ago I went to the Buildathon and made a 7MHz 200mW transmitter. I wanted something more powerful so I have used the wooden base and some of its components to build this new transmitter. It will run at 5W output with a 20V supply. The output filter matches into my ATU Ok; but you may have to experiment with other designs to ensure good harmonic suppression. Soldering on drawing pins is not as easy as it sounds because when adding more components to the same pin the former parts can become loose and move. Note the novel key made with PCB and a wooden spacer.
This simple circuit has featured in a lot of my projects.

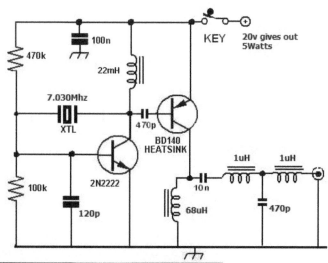

2 Transceivers

Chirpy 10m transceiver
Roger Lapthorn, G3XBM, 37 Spring Close, Burwell, Cambridge, CB25 0HF
http://www.g3xbm.co.uk   http://g3xbm-qrp.blogspot.com/

This is probably the simplest, full break-in 10m CW transceiver possible. Based on my XBM80-2 design for 80m, this is essentially the same circuit redone for a 28.060MHz *fundamental* crystal. Power output is around 100-200mW, which is more than enough to cross the Atlantic on a good day. Unfortunately the keyed oscillator does produce a fair amount of chirp on the CW, hence the name. I've not yet overcome this so I guess one cannot expect perfection in something this simple.

The circuit is a crystal controlled Colpitts oscillator used on TX with the key down and as a direct conversion receiver with a single stage oscillator-mixer and separate audio gain stage with the key up. Very inexpensive, ubiquitous, 2N3904 transistors are used. One can argue whether this receiver is a direct conversion or a regenerative one: in a single stage it is probably not possible to say it is one or the other. At least the receiver is crystal controlled and does not need a reaction control, so on balance it is more likely best described as a direct conversion receiver. The receiver audio output into the high impedance crystal earpiece is low, but it can hear down to around 2uV (-100dBm) in a quiet room. Your ears may be better than mine. This sensitivity has been checked on 3 different examples built. There is no real audio selectivity so you will need to use your "ear-brain filter" to select the 800Hz-1kHz received audio signal.

The frequency shift between RX and TX (about 1kHz) is just about right to listen for replies on the TX frequency. Broadcast breakthrough does not seem to be an issue at all, which is surprising as there is little front end selectivity or rejection of out of band signals.

The transmitter really needs the simple 3 component low pass filter (shown on the RHS in the circuit) adding for serious use, but this was omitted in the basic "no frills" version. It is possible that an ATU will provide the additional filtering in some set-ups.

Notice there is no supply decoupling in the basic version This should be added if the supply impedance is not very low but you can get away without (wishing to keep component count absolutely as low as possible) if using a low impedance battery supply and short leads. Don't try using a mains PSU as audio hum will be a problem.

QRP SCRAPBOOK

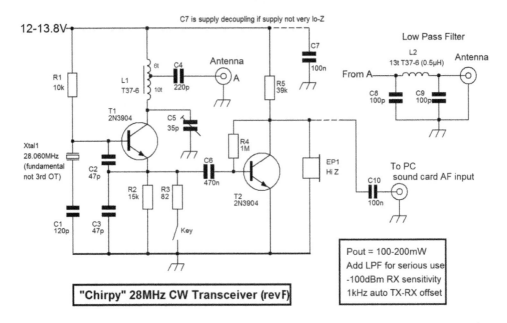

"Chirpy" 28MHz CW Transceiver (revF)

Pout = 100-200mW
Add LPF for serious use
-100dBm RX sensitivity
1kHz auto TX-RX offset

Best DX so far is a QSO with IT9QAU/QRP who gave me RST439. Distance was 1414km. The main issue is the chirp: arguably OK for something this simple, but really not quite good enough. I suspect that with less TX power there will be less chirp.

This is a "for fun" rig, so don't expect incredible performance, but it *does* work. On receive it has copied plenty of stateside and European signals on the QRP calling frequency.

If the fixed capacitor in series with the crystal was made variable some small movement of the frequency would be possible allowing around 10-20kHz of movement around the QRP frequency. This might help get a few more contacts although one has to watch the RX-TX offset does not change too much as the crystal is pulled down in frequency.

By connecting the collector of TR2 to a PC soundcard, a number of simple SDR packages will allow ~28.040-28.080kHz to be monitored. Using the SM6LKM (free) software VLF receiver that tunes 0-22kHz, CW and other signals over a 44kHz section of the band may be heard quite well. Of course, with no I and Q inputs and a direct conversion receiver the two sides of the spectrum either side of 28.060kHz are folded back on each other. Still, this is a simple way of extending the use of this very simple rig, albeit rather negating its ultimate simplicity.

There is more information and a video showing Chirpy in use on my website at http://sites.google.com/site/g3xbmqrp/Home/xbm10_2 .

The 'Tiny Toy' – a 40m CW QRP transceiver
By Peter Parker VK3YE (Originally appeared in Lo-Key, March 2012)

Introduction
You sometimes see circuits for so-called 'minimalist' QRP transceivers. These are typically crystal locked, run a few hundred milliwatts and have a rudimentary receiver.

Such rigs prove that a transceiver needs only a few parts to function. Their intricate construction makes them great conversation pieces. And minimalist rigs are good for teaching the basics; starting simple and adding stages to improve performance is a legitimate design method. As is the opposite - or 'Muntzing'[1] - that removes parts from more complex circuits until they stop working[1].

Minimalist rigs unfortunately often fall short as 'contact getters'. Fixed frequency combined with low power means certain failure, at least in VK. This is because most QRP contacts are made by calling other stations rather than calling CQ yourself. This means you must hunt for other stations on their frequency; they won't come to you. Frequency agility is probably worth at least a 10 or 20dB RF output power increase.

Simple receivers are another weakness. Wide selectivity can perhaps be tolerated as the brain is an effective audio filter. However broadcast station overload, hum, microphonics, poor sensitivity and low volume all make operating frustrating, especially when making the coveted 2 x QRP contact.

The 'Tiny Toy' presented here avoids the worst vices of the ultra-simple rigs. For a start it has a wide-swing twin crystal VXO that covers most of the 40 metre CW segment. And, instead of using the collector/emitter junction of the final transistor as the receiver detector, I've used a separate NE602 stage with some gain. I kept the battery power, but used 4 x AAs for a six volt supply rail. These are cheaper and have longer life than the small nine volt batteries.

Circuit Description
Critical to the rig's success is the wide-swing VXO stage which uses two 7030 kHz crystals in parallel. I achieved 7005 – 7028 kHz coverage with 2 x 15 uH RF chokes in series with it and a plastic tuning capacitor. The crystals are cheaply available from Expanded Spectrum Systems (www.expandedspectrumsystems.com).

The oscillator is followed by a driver and a keyed power amplifier stage. At 6 volts expect around 100 to 200 milliwatts. Those using a 12 volt supply should get nearer 500mW, but I would suggest a 6 volt regulator for the VXO and receiver stages.

The receiver mixer is a standard NE602 stage. I've found a fixed 4.7 uH RF choke and 100pF capacitor to form a satistactory (and small) front end tuned circuit for 7 MHz, but you may wish to experiment with values to optimise.

---

[1] Named after the American TV engineer-salesman who built cheaper and simpler sets suitable for strong signal areas only. The legend goes that he snipped out any part that was not strictly required for the set to work in front of his factory technicians.

QRP SCRAPBOOK

A BC548 audio stage drives a crystal earpiece to good listening volume. An LM386 stage and low impedance phones could be used instead but use more parts and power.

A low pass filter is permanently in the antenna lead. Transmit/receive switching is strictly manual. Change the switch and reset the frequency (no frequency offset being provided). While making the rig awkward for contesting or high speed operating, it aids simplicity. The Tiny Toy will also teach you to how to send Morse without sidetone!

Construction

A shallow U-shaped chassis was formed out of printed circuit board material, with the largest piece being the front panel. This fits into a small food container which also houses the batteries (taped together with no holder).

Solder all components point to point, with grounded parts anchored to the chassis. The NE602 is soldered up side down with the earthed pin (3) carefully bent to touch the board. The faint of heart should use a socket as a damaged NE602 will ruin your day.

Build the VXO and buffer first. Optimising shift may take several hours but is well worth it. I used RF chokes available from Jaycar. Several smaller values in series may work better than one of an equivalent larger value. Short both sections of the tuning capacitor to maximise downward swing and set both trimmers to minimum to maximise top end coverage.

Experiment with inductor and capacitor values as even quite small changes greatly affect coverage. Carefully soldering the two crystal cans to each other and to chassis via a short wire offcut improves mechanical stability and drops the frequency by about 2 kHz. Again don't attempt this unless confident. Aim to get the maximum VXO pulling range consistent with stability and output; the more you pull a crystal the greater the risk of instability and/or chirp.

The transmitter PA and pi network can be built next. With the key down the output should sound clean without broadband hash or radiation on other frequencies. If this is not the case, tame the beast by experimenting with decoupling capacitors, ferrite beads, low value base resistors and the like.

Once happy, build the receiver. This should work first time. At least in urban areas sensitivity should be sufficient to hear band noise. If your audio output is low, try another crystal earpiece as I've found that they sometimes vary. Failing that (or if you don't have a crystal earpiece) use an LM386 stage instead.

In use the main differences between this and other rigs is (a) the lack of a sidetone, (b) manual transmit receive switching, and (c) the need to zero beat the other station before transmitting (there being no automatic 800 Hz offset). This takes a bit of getting used to and makes the rig less suitable for fast contest work.

More advanced builders may care to add automatic T/R switching and frequency offset. Obtaining a constant shift is difficult with VXO circuits. One possibility, successfully used by the author in another project, is to have two variable capacitors on the front panel, with the T/R switch or relay changing between them. This provides split frequency capability and the ability to independently tune the direct conversion receiver off to whichever side provides clearest reception.

Results

A couple of hours after dawn or before dusk seems to be good times to operate with milliwatts on 40 metres. Band noise and competition from DX signals is less than during the hours of darkness.

Mere milliwatts can be heard several hundred kilometres away. The first contact was with Tom VK7LF, followed by Mike VK2IG and Keith VK2WQ. Contact with VK2 (spanning distances up to 700km) was repeated the following weekend, this time from the beach. The antenna was a 22 metre end-fed wire, tuned by a small L-match coupler.

Conclusion

While its power still puts it in the novelty rig class, the Tiny Toy's wide frequency pulling range and reasonable receiver make it more usable than many. It's no DX machine but fun to build and use, with some amazing distances possible.

Tiny Toy on YouTube
ww.youtube.com/watch?v=DvHfRJvjUN4

QRP SCRAPBOOK

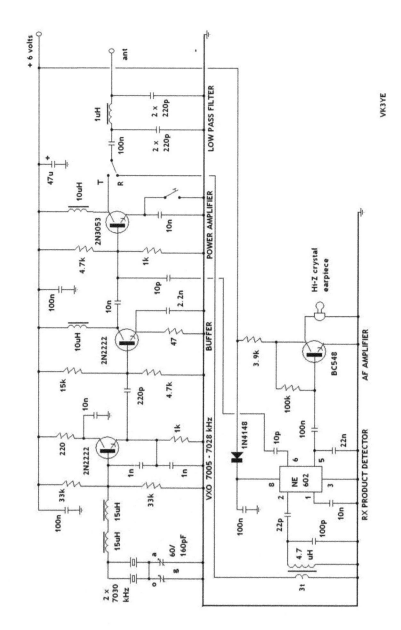

'Tiny Toy' 40 metre CW transceiver — VK3YE

PART 2 TRANSCEIVERS

The Bigger Toy QRP Transceiver
By Peter Parker VK3YE (first appeared in Lo-Key September 2012)

The 'Tiny Toy' featured in SPRAT 151 was a practical minimum parts pocket sized QRP station. It produced some good interstate contacts and was a great conversation piece.

However, as is the case with all simple rigs, a few more components will overcome many small-rig compromises. You'll span longer distances more often and get more readability 5 reports. And any extra cost involved is a bargain if it doubles the contacts made.

Featured here is the Tiny Toy's larger and more capable brother.
It has eight transistors versus the Tiny Toy's four. However there are no ICs and you don't need the dexterity of a jeweller to build it. All parts (except for the crystal) are available from Altronics or Jaycar. For these reasons plus the better performance, I'd recommend this over the 'Tiny Toy' if constructing a first or only QRP rig.

The rig started on 30 metres. Good reports were received but activity, especially from within VK is lower. The hint was taken and it was moved to 40 metres, which is recommended for a first QRP rig. 80 metres should also be possible but has not been tried.

Circuit description

Where are the extra parts?

The VXO has an extra tuning capacitor switched by relay. Two capacitors provide split frequency operation and any amount of transmit/receive frequency offset. It also allows QRM to be dodged, as unlike a fixed frequency offset, you can listen on either side of the local oscillator's frequency. It's as good as having two VFOs and I don't know why more QRP builders don't add this feature.

There's better buffering. That's why the transmitter has five transistors versus three. The buffer reduces loading on the oscillator when the rig is keyed. A fixed crystal oscillator can do without a buffer, but a VFO or VXO needs one to prevent chirp. The buffer's two dollars worth of parts is a small price to pay for better stability and tone.

Some extra output power comes in handy when your signal is often near others' noise levels. One watt as against 100 milliwatts makes a big difference. Doubling supply rail voltage from 6 to 12 – 15 volts was a big help and allowed use of common power and battery supplies.

The receiver also had a makeover. Discrete transistors (instead of the NE602) improved performance, lowered cost and made the project more 'junkbox friendly'. The double balanced diode mixer provides strong front-end performance and better resistance to AM broadcast station detection than the two diode single balanced mixers often seen. A high-gain two transistor amplifier capably drives a sensitive crystal earpiece. The receiver's only other transistor boosts the output of the VXO to drive the diode product detector. The receiver's performance is excellent and you'll find it free from the hum, microphonics

QRP SCRAPBOOK

and overload that can plague other simple designs. Gain is sufficient not to require RF preamplification, at least up to 10 MHz and it would make a good 'bolt-on' module for CW or DSB transmitters.

This is not a high-speed break-in DX rig. For this reason I went with a simple DPDT transmit/receive switch. One section changes the antenna while the other switches between transmit and receive frequencies. The main improvement over the 'Tiny Toy' is that frequency offset is both automatic and infinitely variable.

Construction

The absence of ICs and the larger box allow less cramped construction than in the 'Tiny Toy'. All parts can be mounted dead bug style on blank circuit board material, without the need for a punched board for ICs. Extensive use is made of pre-wound RF chokes; with the only winding being around the broadband product detector and the transmitter PA stage.

I started with the VXO and buffer. This is easy to get oscillating but requires a lot of 'cut and try' to get the pulling range right. On the one hand coverage must be sufficient. If not, the station you wish to work will always be just outside it and won't hear your call. But on the other coverage below 7 MHz is wasted and reduces stability.

I suggest starting with a few microhenries in series with the crystals, measuring the pulling range and gradually increasing it by wiring another choke in series. Keep substituting until an adequate range is obtained. This can be as much as 30 kHz with two 7030 kHz crystals such as obtained from Expanded Spectrum Systems. Altronics sell RF chokes cheaply and leftovers will be useful for other projects.

The audio amplifier is easy to build. Testing is easy – apply power, connect earpiece and touch the input connection to hear a noise in the earpiece. I haven't tried magnetic phones with this circuit but if you do use a series coupling capacitor. An LM386 or similar stage could also be connected here to drive a speaker.

Though it's the simplest stage, the receiver product detector takes the longest due to the winding of its two bifilar transformers. Even this is fairly simple, with only a hand drill, pliers and a multimeter to test being required. Wire is about 0.3 mm diameter (not critical) and can come from old transformers. However diode and winding polarity is critical.

The driver and final amplifier stages need to stably amplify the buffer's output to approximately one watt. Hash and instability with the prototype was cured by adding the 47uF from the driver's emitter. The key shorts both driver and power amplifier to earth, meaning significant current goes through it. If this is a problem (eg if using a delicate microswitch as a key) it may be worth adding a keying transistor to handle the current.

Like in the receiver product detector, the PA collector inductor is wound on a 2 hole ferrite TV balun former, but with two windings instead of three. The pi network uses RF chokes for the coils. Polystyrene capacitors are preferred for its capacitors though disc ceramics can be used with some loss.

PART 2 TRANSCEIVERS

Afterthoughts

The rig has provided solid contacts around South-east Australia. Its one watt can be expected to occasionally reach to New Zealand, though for these distances five watts is much more reliable.

While its basic RF performance beats the 'Tiny Toy', it still omits a few operating niceities. These include automatic transmit/receive switching, sidetone and low keying current. Also where there is extraneous noise two earphones are better than one earpiece. Either try another in parallel or modify the audio stage for low impedence phones.

Bigger Toy Front View

Bigger Toy Inside view

Bigger Toy 40m QRP CW transceiver

Modifications to the VK3YE Bigger Toy 40m Transceiver
Duncan Walters G4DFV. duncan.walters12@ntlworld.com

Having built the original VK3YE transceiver (SPRAT 153), the following modifications were carried out in order to provide more RF output and to change the keying arrangement. By making some circuit alterations and using a power MOSFET in the P.A. stage, significantly more output was achieved.

On my prototype design, with a DC standing (quiescent) bias current set to 5mA, an RF output of 4 Watts was achieved as measured using a 50 Ohm dummy load wattmeter.

All the existing VXO, and buffer stages were kept as original, the 2N2222 driver was replaced with a 2N3053 with different base biasing resistors 1k and 4k7.
A 22 Ohm resistor was inserted in the emitter. The 2N3053 was fitted with a clip-on heatsink. The P.A. stage is left unkeyed, only the driver stage is keyed.
During key-up, the P.A. stage draws 5mA, on key-down, a drain current of around 490mA is produced. (With a stabilised 13.8v supply).

The P.A. stage was replaced by an IRF510, and the standing (quiescent) bias adjustment is by a 10k preset. The output from the P.A. stage is via a 1n capacitor
followed by a conventional lowpass filter to the antenna switch.
The IRF510 was fitted with a TO220 style finned heatsink.
As the original 1uH chokes in the low pass filter would likely saturate at the higher power level, these were replaced by inductors wound on T37-2 toroid cores.
The original bifilar wound BD139 collector choke was replaced with a single winding of 10 turns on a T68-2 toroid core in the IRF510 drain..

To set the standing (quiescent) bias current, before applying power, ensure that the slider of the 10k preset bias pot is set to the 0v end of its travel.

Open the connection at point X as shown on the circuit diagram.
Insert a DC ammeter (or multimeter set to 20mA DC range)

Ensure the key is up or disconnected and apply 13.8v supply.
Observing the meter, turn the slider of the bias pot slowly until it shows a steady 5mA. Be careful not to go above 10mA as the IRF510 may start to get hot.

Switch off the supply and restore connection at point X.

Connect a key, and re-apply power. Select "Transmit" position with the T/R toggle switch (as per the original design). With key down, check transmitter output using whatever reliable method you have to hand.

You should be able to achieve around 4 Watts into 50 ohms.

QRP SCRAPBOOK

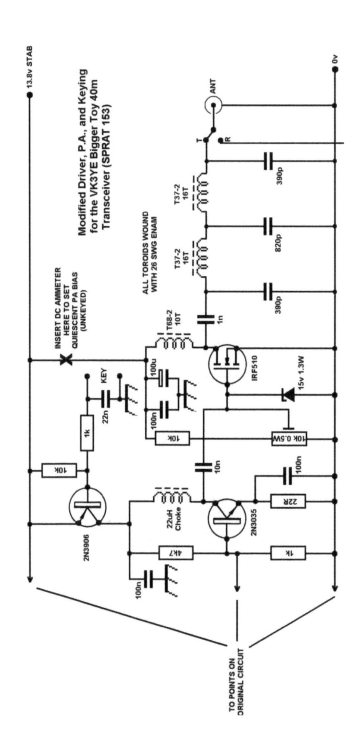

PART 2 TRANSCEIVERS

> # The 3560 Transceiver
> Peter Howard, G4UMB, 63 West Bradford Rd. Waddington, Lancs

This transceiver was made to fulfil two needs. 1. To make a 4 Watt transmitter to give a decent output to ensure I got some contacts. 2. To make a receiver to go with it which is simple yet sensitive enough. I tried to use the same value components whenever possible because my supplier will only often deal in packs of ten. The DPDT switch can be changed for a relay to make a break in change-over. In use the receiver is easily overloaded with night conditions and can be susceptible to continental broadcast interference. I made the LM386 ic AF stage very sensitive because you need good gain during the day; but at night it's too much and distortion happens. So there are plenty of improvements that can be done. It was made on strip board. Another disadvantage of such a simple transceiver is that it only works on one frequency of 3.560 MHz I chose this because the GQRP club sells the crystals at £2 each making them economical to buy 2 to save extra switching. Perhaps two people could get together and make two transceivers and work each other with them. The trimmer capacitor is adjusted to offset the transmitter frequency to produce the audio tone as it's a direct conversion receiver. I did the adjusting and setup using my main station rig. All the coils were bought as chokes. With such a simple Pi filter on the input/output you will need to adjust the values of the 820p caps a little to get a good match. I have not fitted a heatsink to my IRF510 yet; but it might be wise to do so because my rig has a current of 800ma at 20v on transmit.

QRP SCRAPBOOK

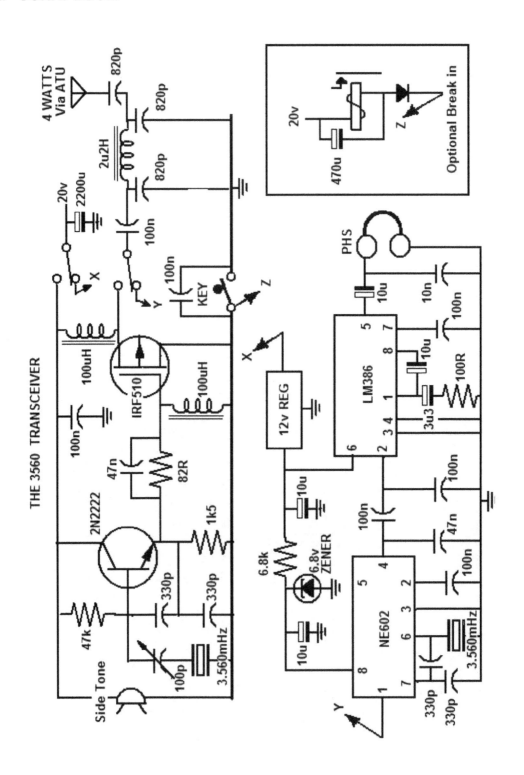

PART 2 TRANSCEIVERS

Chinese X1M QRP Transceiver
Steve Farthing G0XAR, Jan Verduyn G5BBL and Alan Rowe M0PUB

In 2013 a diminutive Chinese made QRP transceiver, the X1-M was announced in the USA. At the bargain price of around $250 the radio boasted an output of 5 watts CW or SSB on the 80, 40, 20, 15 and 10 meter bands, general coverage receive and, the option of transmitting out of band without a low pass filter. All this in a sturdy case measuring 16x10x4.5 cm and weighing in at 600 grams, including a serial interface for computer control and a microphone, it just had to be the bargain of the year. So, as we could not buy one in the UK we ordered one via a fellow amateur in the USA. However this was not without problems as thanks to an incompetent courier who did not deliver it and sent it back, various bouts of illness and other things it was not until a year later that it arrived.

So having had the radio for a few months now this is our "warts and all" experience of the X1M, serial number 59, one of the first Chinese HF transceivers to come to the Amateur market.

Our first impressions were not very encouraging. The case had clearly been drilled by hand and the packaging was inadequate. the 9 pin computer interface cable, required for computer control and firmware upgrades had been damaged during transit. The manual was a photocopy. We felt that the radio had been home made rather than factory built. Also during testing we discovered some poor soldering on the microphone and the front panel buttons. But we are QRPers well used to making our own radios and we enjoy a challenge. And most of our radios have been "home built"!

Operating the radio, in common with a lot of more expensive sets, requires the use of front panel controls which talk to a microprocessor with readout on a "white on black, 30mm x 16mm LCD screen, which while small and crowded with information, was perfectly readable to all of us, two of who are on the wrong side of 60. However if you have a vision impairment this might not be the radio for you. However you do have the option of operating the radio entirely under computer control should you wish,

The front panel has an AF gain control, and a tuning knob whose rate can be varied from 1Hz to 10 MHz by toggling front panel buttons. The LCD displays all the settings such as mode, frequency, RIT which can be changed by using the front panel buttons one of which can lock the settings.

There is a 3.5mm socket at the front for the supplied microphone, on the back there is a BNC socket for the aerial, our radio came with a slightly deformed one but it worked. Also 3.5 mm sockets for a key, and headphones, an 9 pin CAT connector for computer control and firmware upgrades, and a 5.5 x 2.1 mm coaxial power connector (the one from an Elecraft K2 fits). Internally there are presets for CW sidetone level and RF output for SSB.

QRP SCRAPBOOK

The RX is quite sensitive. There is no AGC as standard however there is a kit available to add this. The provision of AGC is a matter of personal taste, and we did not see the lack of it to be a major disadvantage having grown up on radios without it.

ON CW there is no way of altering output power, with SSB it is possible to reduce power by lowering your voice level or distance from the microphone. As there is no ALC or AF limiting provided it is a case of "suck it and see" with your voice level and distance from the mic.

Frequency readout is pretty accurate and SSB selectivity is adequate however strong signals will be heard when tuned to the unwanted sideband.

Jan and Alan carried out some performance testing with the following results :-

Band	Sensitivity uV 10dB Sinad preamp on	Preamp off	RF Output Power Watt CW	Current Consumption Amps
1.820	0.27	0.72	Rx Only	RX 0.51/0.47
3.600	0.27	0.73	5.5	1.46
7.100	0.35	0.8	5.4	1.37
10.100	0.28	0.51	RX Only	RX 0.51/0.47
14.100	0.3	0.67	4.0	1.3
18.100	0.25	0.47	RX Only	RX 0.51/0.47
21.100	0.25	0.6	3.8	1.25
25.100	0.27	0.6	RX Only	RX 0.51/0.47
28.100	0.45	1.0	2.7	1.1

We also discovered that it is vital to keep the firmware (the program that runs in the microprocessor inside the X1-M) up to date. Ours came with a very early version which had some problems with the internal keyer, dashes were only 2 times the dot length instead of three. However later versions of the firmware solved this problem. To update the firmware you need to buy the optional cable and, unless your computer has a serial port, a USB to serial convertor. EBay is your friend here.

PART 2 TRANSCEIVERS

After a few months of evaluation and some QSO's in SSB and CW what do we think about this tiny newcomer? Initially there were some teething troubles. but nothing we could not overcome and given the price the radio is pretty good value. It performs acceptably and is reliable. The internal keyed is tricky to use because of varying latency but there is no problem for the straight key operator. The bandwidth in CW mode is considerably wider than a more expensive radio, say the Elecraft K2 at identical output levels, because there is no shaping of the characters whatsoever so keyclicks are clearly audible +/- 6 kHz away from the TX frequency.

Transmission outside of the ham bands, and in the WARC bands is possible, however there is no LPF provision for these frequencies. So if you want to do this you have to make your own. Being able to transmit outside a legally authorised band is a thorny subject. But it could be useful for transvertor users. And also the LOWFER community as coverage starts at 0.1 MHz.

We are concerned at the high current consumption on receive/standby. Half an ampere will soon drain those batteries so we would not recommend it for hikers or SOTA use.

Time moves on and the X1M has been subject to an upgrade and improvement. It is now replaced by the X1M platinum edition and is carried by some suppliers in Europe. As there is no EU type approval this might be in kit form, best to check before hand.

Lastly we would like to say thanks to Paul Maciel AK1P for helping us get the set, Fred Lesnick, VE3FAL, Paul Ross W3FIS and Charlie Vest W5COV for the English User Guide. Members of the Yahoo X1M user group who have answered our questions and been responsible for the improvement of the radio, especially with the firmware.

References :-

1. X1M User manual by VE3FAZ, W3FIS and W5COV
   http://www.qsl.net/ea3gcy/x1m_archivos/X1M%20manual%20english.pdf
2. The X1M_QRP-Transceiver newsgroup at yahoo.com for user experience, schematics, AGC kit info and firmware upgrade files

The Foxx with relay QSK
Peter Howard G4UMB

The FOXX TX/RX has been around for some time and I have found it to be a good QRP set. The one I have built here adds a simple QSK (Auto change over) by holding a relay on a second or so after keying finishes. So a 4 pole changeover relay does the switching. I wanted to make it suitable for the 80M 60M 40M bands as the Club sells the Xtls 3.560Mhz, 5.262Mhz and 7.030Mhz. The output at 18v in is 4 Watts. Note that on keying the voltage to the xtl osc increases making it more powerful to drive the BD140. If you wish you can use the spare switch on the phones to wire in a side tone circuit taking the positive feed from the key.

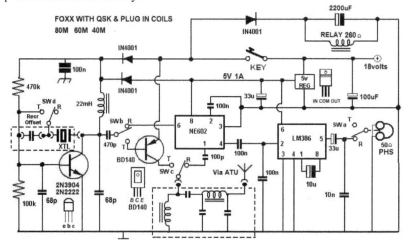

In order to reduce the amount of switching I built each tuned circuit on a plugin board. The xtl and offset capacitor are also plugged in a socket as are the transistors and IC's making it easily serviceable in case of a component failure. You can try a 10pF on 80M as an offset Cap and a wire link for the other bands. I used a stripboard PCB. An extra resistor across the relay coil could reduce the delay on switchover. With such a simple tuned in and out circuit a good ATU is a must. As the picture shows, I have mounted it cheaply on a wooden base unboxed, unscreened but results have been satisfactory.

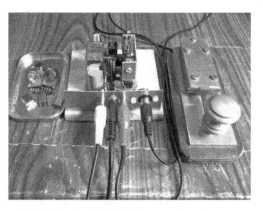

3 Receivers

40m CW Receiver
Barry Zaruki M0DGQ, 26 Heathfield Rd, BIRMINGHAM. B14 7DB

Here is a 40m CW receiver for use with the 40m CW QRP transmitter (SPRAT 151). Again cheap, easy to obtain components are used throughout the design. Performance is good, signals below 2uV at the antenna socket are easily copied and selectivity is excellent due to the use of a 9MHz CW crystal filter (G-QRP club). The set also has RF derived AGC although this can be omitted if desired reducing the component count slightly. Local oscillator signal injection is provided by a VXO using a 2MHz ceramic resonator giving coverage of most of the CW portion of the 40m band. Good quality VFO's can sometimes be difficult to achieve for those not experienced in building these, so a VXO is used here. The performance of the VXO very good, after ten minutes "warm up " it is absolutely solid, no tuning adjustments need be made for half an hour or so, I was pleasantly surprised by this considering a cheap polyvaricon tuning capacitor is used.

Circuit description
The antenna signal enters the set via a two pole bandpass filter L1, L2 feeding Tr1 a FET common source RF preamp. Tr1 provides little loading on L2 as its gate is high impedance thus maintaining the Q of the filter. The FET preamp is used to overcome the insertion loss of the mixer and crystal filter. The drain load for Tr1 is T1 which along with D1, D2 form a single balanced mixer.

Local oscillator from the VXO is injected at the centre tap of T1. From here the mixing products pass through T2, 9:1 impedance ratio transformer in order to provide a good match to the 50 Ohm impedance crystal filter. The wanted IF signal leaves the crystal filter via a 1:9 step up transformer, this provides a good match to following cascode IF amplifier. Two cascode IF amplifiers are used giving a gain of roughly 70dB in total. The IF transformers used are standard 10.7 MHz interstage IFT's with an additional 10pF connected across the primary. The second IF amp feeds a diode product detector D3, D4. and also feeds the AGC amplifier.

A crystal oscillator, Tr8, is used as a BFO for carrier re-insertion at the product detector. A 9.0015 MHz or 9.000 MHz xtal is used here as the CW filter has a centre frequency of 9.0008 MHz, this is probably so a 9.0015 MHz USB Xtal could be used for carrier reinsertion in the set it was intended to be used in thus saving the cost of a separate CW carrier Xtal. (several of these filters were sweep tested and they all had a centre frequency of 9.0008 MHz). Recovered audio from the detector passes to common emitter preamp Tr9 via the volume control and then to the audio power amp. A LM386 can be used here (you will still need the preamp Tr9) if you wish,
indeed it is probably cheaper to do so. I used the discrete audio amp as a design exercise and to keep the set "chip free ".

VXO
The circuit for the VXO is straight forward. A Colpitts variable oscillator is used. A zener diode is used to stabilise the DC supply to VXO Tr6. A buffer Tr7 follows the oscillator

QRP SCRAPBOOK

which reduces oscillator loading and pulling by the mixer. A cheap polyvaricon variable capacitor (G-QRP) is used for the tuning control, no reduction gearing is required - tuning is smooth and easy providing a large control knob is used. Approximately 40 kHz of swing is given by this VXO, resulting in a band coverage of 7.00 MHz to 7.045 MHz thus matching the range of crystals used in the transmitter. Experimentation with the feedback capacitors in the VXO may give a greater tuning range. Polystyrene feedback capacitors are to be recommended for use in the VXO, I used a mixture of polystyrene and ceramic disc as I did not have correct values available. If you are an experienced VFO builder then you could build a VFO giving coverage of the whole of the 40m band if desired.

AGC amplifier

RF derived AGC is used in this set. A small proportion of the IF signal is taped off from Tr5 collector via C1 a 3.3pF capacitor to feed an emitter follower buffer Tr10, therefore very little loading is presented to Tr5 collector. A low gain preamplifier Tr11 follows the buffer via the AGC threshold control P1. Tr11 collector load R7 feeds a rectifier / voltage doubler D1, D2. The result is a DC voltage proportional to the received signal strength present at the cathode of D2. This is used to control Tr12, the stronger the received signal the harder Tr13 conducts thereby reducing the voltage on the AGC control line. During TX the AGC voltage is grounded by Tr13 allowing P2 control of the IF gain for correct sidetone level. It is important to keep any BFO signal out of the IF amplifier (apart from the IFT2) as this will upset the AGC action, screening should be used between the BFO and IF amp. The main purpose of this AGC circuit is to stop strong stations from deafening you when the IF gain control is at maximum. If you do not require AGC then it is a simple matter to use a pot as a manual gain control, details are shown for this in the AGC amp circuit diagram. No side tone oscillator is required as the receiver will pick up the TX due to proximity; P2 is adjusted for a suitable sidetone level whilst holding the key down.

Front Panel showing Main board and VXO board

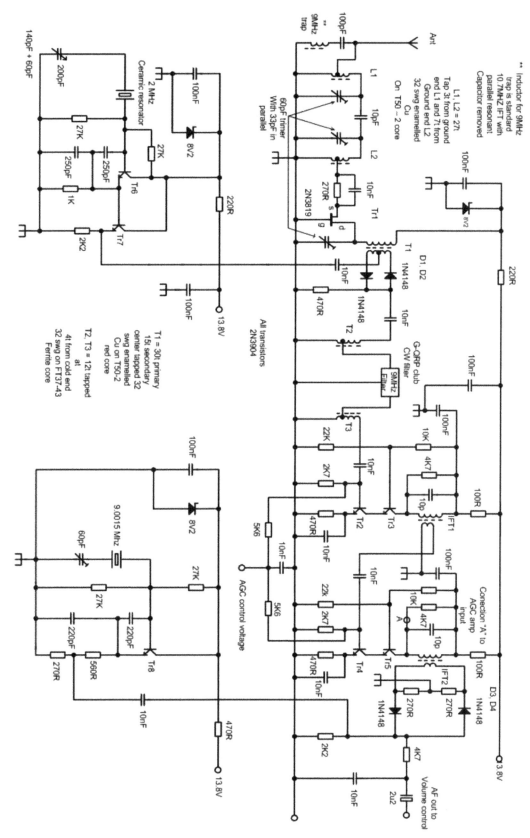

QRP SCRAPBOOK

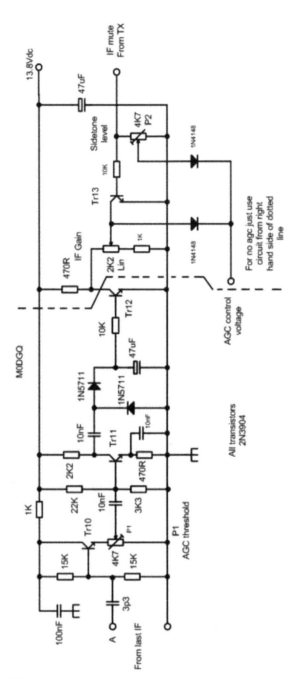

AGC Amplifier

PART 3 RECEIVERS

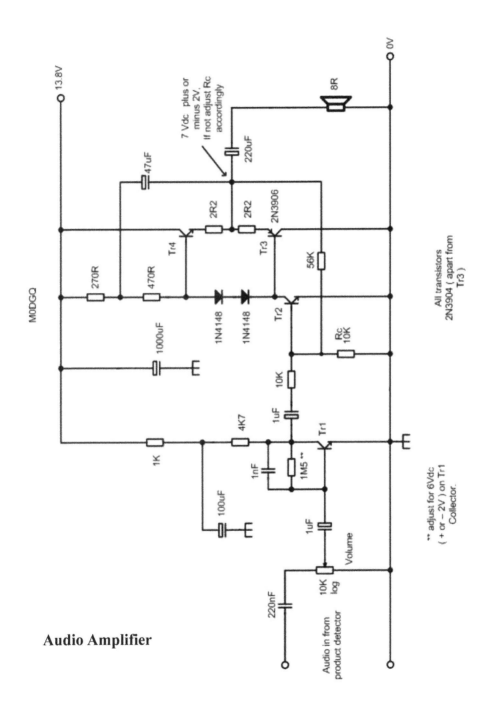

Audio Amplifier

QRP SCRAPBOOK

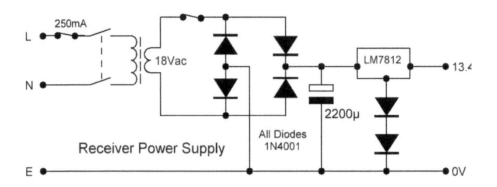

Receiver Power Supply

Complete Board

Inside view

PART 3 RECEIVERS

The G4DFV / G4GDR 80m One Valver
Rev. A. Heath, G4GDR, 227 Windrush, Highworth, Swindon, SN6 7EB

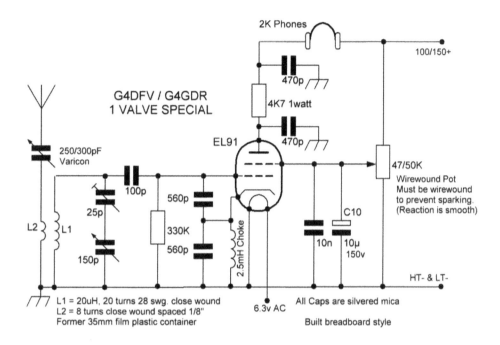

I claim no originality for this simple 80m receiver built as a "together project" with Duncan, G4DFV. It has been my desire for some time to put together a simple one valve receiver and a one valve transmitter; emulating a 1920's station. The receiver is rock steady once tuned onto a station. Switching off the HT to transmit and then on again after transmission shows no drift. This is unlike most regen receivers. The secret is the EL91 audio valve which is far better for regen receivers than RF valves.

The HT is 150 volts run from a stabilised line with the output to high resistance phones. However, I use a small transistor amplifier which switches speaker or phones and gives a greater audio output. It is really a great little receiver that receives SSB and CW very well. Using the components shown, mine covers the whole 80m band. I use a velvet vernier drive from a TU5B and no band-spread is required.

The polyvaricon capacitor in the antenna input could by a fixed value of around 150-200pF (experiment). Adrian is building a 40m version and Duncan has made coils for 160, 80 and 40m and used a second EL91 as an audio output stage.

This little receiver and a simple valve transmitter have worked all around Europe.

The "Twelvevolter" – A Hybrid Receiver for 40m
Duncan Walters G4DFV, 11 King George Fifth Ave. Mansfield. NG13 4ER

This receiver utilises both valve and solid-state technology, but the unique feature of the set is it uses just 12 volts DC for the solid-state circuitry as well as providing heater and HT supply for the valves.

Operating valves at low HT is certainly is not something new, but to achieve useful gain in the valve stages of this receiver, I had to test quite a number of valves before I got results. In fact some valves just utterly refused to work at all with such low HT.

The best valves I found fit for my purpose were the E88CC double triode and the EF95 (6AK6) miniature pentode. The EF92 also works too, but has a slightly different pinout.

One half of the E88CC is used as a grounded-grid RF amplifier, with the antenna being capacity coupled into the cathode via the 1K attenuator pot. The signal is coupled into the main tuning tank circuit comprising L1 and the associated capacitors. 40m was chosen as it is my favourite band for listening, as there is usually always some signals to be heard regardless of what time of day it is.

A useful phenomenon I discovered is that the RF attenuator pot also acts as a very fine tuning control, particularly useful when resolving SSB signals.

The former for L1 was a 50mm length of 15mm diameter plastic plumbing "barrier" pipe which I originally obtained from my local DIY centre. A 10mm length of wooden dowel, drilled in the centre, was superglued into the end of the former to allow it to be screwed down to the chassis.

The 60p trimmer across the 68pF capacitor across L1 is adjusted to get the tuning into the right "ball park", whilst the 30p trimmer in series with the 100p main tuning capacitor is adjusted to set the required bandspread.

The EF95 is configured as a regenerative detector, regeneration is controlled by screen grid voltage set by the 47K linear pot.

Detected audio is coupled into the control grid of the second half of the E88CC. The amplified audio signal is developed across the primary of an LT44 audio transformer.

The E88CC and EF95 heaters are wired in series across the 12 volt supply, but as the EF95 only requires 0.175 amps of current, it is shunted with a 47 ohm 2 watt resistor.

PART 3 RECEIVERS

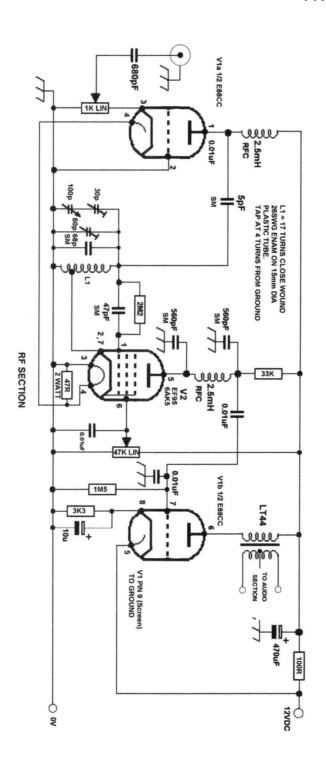

QRP SCRAPBOOK

Incorporated into the receiver is a passive SSB and CW filter. I found this very effective at reducing background noise in both modes.
The filter switch has three positions, OFF, SSB, CW.

Each section has 4.7K preset level controls. These are used to set roughly the same audio level from each of the three selected switch settings as heard in the speaker. The 122nf and 147nF capacitors used in the SSB section of the passive audio filter are made up from 100nF and 22nF in parallel and 100nF and 47nF in parallel. I use mainly polyester capacitors in the filters with tantalums for the 1uF and 2u2F values.

The audio filter and audio output amplifier were built on single-sided PCB boards. No artwork was used, I simply drew the tracks with a Dalo pen then etched and drilled to take the components.
The completed boards were mounted vertically on each side of an L-shaped piece of aluminium sheet. This was bolted on to the top of the main chassis.

The secondary of the LT44 transformer is connected to a preamplifier stage before the audio is passed into the input of the audio filter.
After filtering, the signals then pass into the main audio stages where they are amplified sufficiently to drive a loudspeaker. The choice to use solid-state amplification instead of valve circuits is that it is quite difficult at low HT to derive any useful audio gain from them. This is one reason that the vintage car radios fitted with valves had a power transistor in the output stage in order to get sufficient gain to drive the speaker.

I claim no originality for the receiver and audio amplifier circuits, the whole receiver was built up of bits from other designs.

PART 3 RECEIVERS

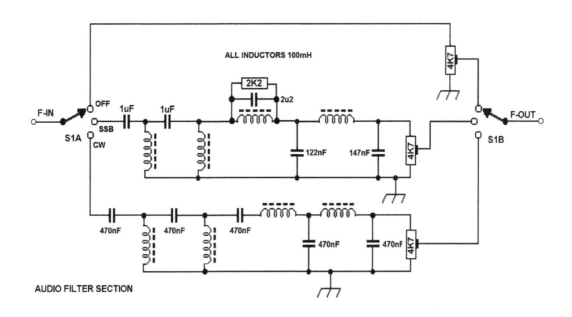

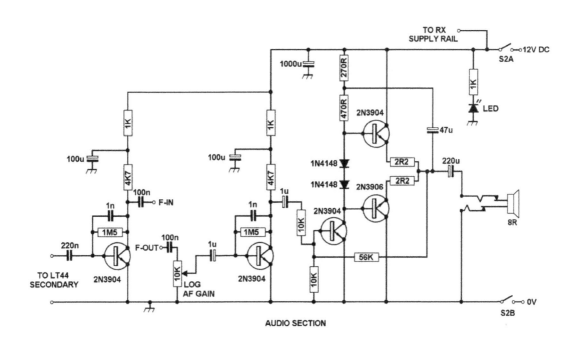

QRP SCRAPBOOK

Sudden PSK?
Steve Hartley, G0FUW, g0fuw@tiscali.co.uk

The Bath Buildathon Crew have been running construction events in Bath and helped out at the Club Convention at Rishworth for about 5 years now. We were asked to staff a Buildathon at the RSGB's Centenary Day at Bletchley with a project that would be attractive to youngsters. We thought about something in keeping with the venue, a spy set maybe, but in the end we went for a data receiver that could be used with a computer; something to reflect the more modern face of amateur radio and young folk love screens. The full details of the project are to be published in the September 2013 RadCom magazine together with opportunities to buy PCBs, crystals, etc, but I thought SPRAT readers might like to see the circuit and maybe have a go at building from scratch, or even by modifying a Club kit.

The idea came from Dave Benson's Warbler transceiver that used an 80m crystal filter ahead of a direct conversion receiver. That circuit did not work out for us and we dabbled with a circuit from the on-line magazine 'Nuts & Volts'. That led us to the conclusion that we were actually looking at a modified Sudden receiver, a circuit we had used in previous Buildathons. We modified the front end with a crystal filter, added an RF amp, and bingo! We added a small gain control on the LM386, but it is not critical. Club Sales have the crystals and the front end coil/FET parts for 80m. With the circuit below you could buy a Club Sudden 80m receiver kit and have it receiving PSK31 in no time; just trim the oscillator to 3.581MH, plug into the PC mic socket, run the software and peak the front end.

The RSGB decided that the Centenary Day project should be on 20m to demonstrate the international dimension of the hobby, so modifications had to be carried out. I tried a prototype with 14.060MHz crystals and soon had a neat little single channel CW receiver. The PSK frequency crystals (14.070MHz) were not readily available and a bespoke order was despatched to the Far East (I understand these will be available from the RSGB if you want to build a 20m version). It is still a modified Sudden and due credit was given in the construction manual, which should be available for download by the time you read this.

To decode the PSK signals we have used a Raspberry Pi running FLdigi, a laptop running Digipan, a PC running HRD/Digital Master 780 and a smart phone running a PSK app.

Twenty one receivers were built on the day and all worked, once a few solder bridges had been cleaned up, and the school children loved it (so did their science teachers). The youngsters that attended the Centenary Day were also given a Raspberry Pi computer each by RS Components. You can see a video of the Centenary Day on the RSGB website and that includes some good footage of the Buildathon.

The circuit diagram here is for the 80m version. For 20m the following component changes are required:

Crystals 1 to 3 = 14.070MHz (trim to 14.071MHz)
Capacitors C3, C4 & C7 = 10pF
Coil 1 = Spectrum 10K $2\mu 6$H

PART 3 RECEIVERS

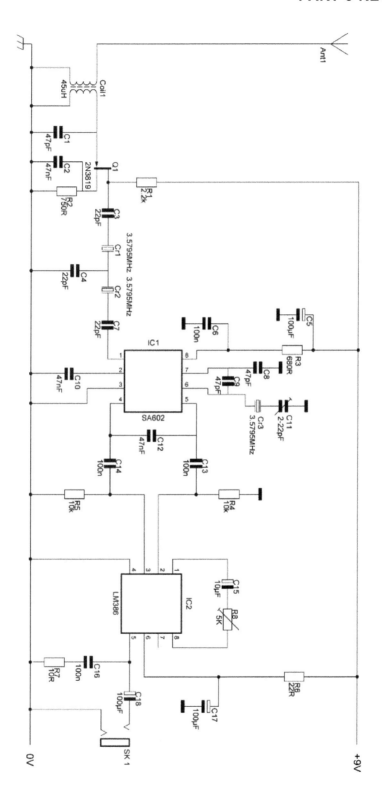

QRP SCRAPBOOK

W1FB MEMORIAL ENTRY
The Mousetrap Receiver
Johnny Apell SM7UCZ, Ekedalsvägen 11, S-373 00 Jämjö. Sweden

The Mousetrap is an SMD project... But on wood with drawing pins! It was a test to see how simple it is to build a CW/SSB receiver with 3 transistors (one Darlington) and one coil. To keep the costs down, the receiver uses permeability tuning with a moveable single turn of Ø1.7mm copper wire from 2,5mm² installation cable. The single turn forms a mouse trap moving in relation to one coil, wounded on 25mm broom shaft and fashioned as a mouse's head! The permeability tuning is based on an idea by PA2OHH. The oscillator is running at half frequency, and is doubled in the mixer by the 2 diodes following the popular RA3AAE mixer doubler circuit.

Naturally, BC station can be a problem with the lack of input tuning; there is room for for improvements... but it is only meant to be a basic design. Output is enough for 32R headset. Most of the gain is from the Darlington pair. I build my Darlington pair from 2×BC550C with the power gain from the last transistor.

Above: The Mousetrap Receiver showing the "SMT" construction using drawing pins

Left: The mousetrap and mouse permeability tuning. A lever on the "trap" does the tuning. A slight kink in the wire going through the wood maintains the tuning position.

PART 3 RECEIVERS

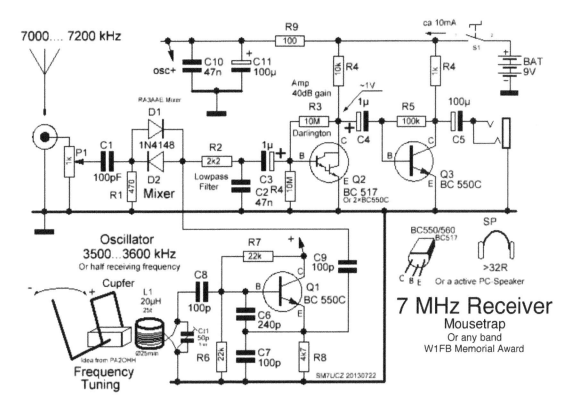

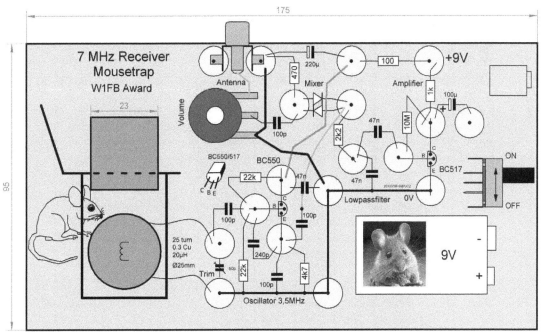

7 MHz Receiver
Mousetrap
Or any band
W1FB Memorial Award

QRP SCRAPBOOK

The RR9 Short-wave Receiver
Rev. Keith Ranger, G0KJK. keithcath@ranger144.fsnet.co.uk

For all too long now I have been QRT! My XYL Catherine and I are living in temporary accommodation owing to urgent house repairs but should shortly be moving back. No outside aerials can be erected here so I have been an SWL during recent months using self-built direct conversion or regenerative receivers and a one metre- long desk-mounted telescopic aerial. The RFI in this block of flats is horrendous but good results on nine amateur bands have proved available on the design to be described. I offer it to fellow-experimenters in the hope that it will give hours of listening pleasure at relatively low construction cost, both in terms of time and money.

What is perhaps slightly unusual about this particular home-brew regenerative receiver is that it covers all the major amateur bands, both CW and SSB, from 80 to 6 metres. Many published designs cover just one, or from one to three, bands but these almost always do not include 6 metres. In any case, restriction to just three or so bands can be frustrating if conditions are unfavourable, one needs versatility. The satisfying thing about the VHF 50MHZ band, with all its varied propagation possibilities, is that just 4 turns of insulated wire on a T-50-6 toroid core placed in the tuning circuitry are necessary for its coverage by this receiver! I have left out Top Band as I seldom hear much activity on it and have gone for the other end of the spectrum instead!

I must emphasise that you must be prepared to "cut and try" to get accurate band coverage but the circuit details I give represent a sound starting-point. Very probably you'll hit most bands first time with the inductor and/or capacitor values specified. However, stray capacitances from the layout style you adopt could change coverage a bit but it should not be by all that much. Keep at it and you will win! For these reasons no point to point wiring instructions are given, especially as I am rather old fashioned and prefer using tag strips rather than printed circuit or perf boards.

I am assuming that G QRP Club builders of this circuit will not be making it a first project because you will already have some basic construction experience and be aware of the principles that need attention, especially the necessity of keeping RF and AF wiring well apart to avoid annoying instability problems. My prototype is stable, selective and sensitive and even on 6 metres there are virtually no hand capacitance effects. This problem is avoided by using just a loosely coupled telescopic aerial but the sensitivity of the circuit is reflected by the powerful SSB signal received from VK3MO on 20 metres and some eleven entities logged to date on 6 metres SSB or CW including OH, EA, S5 and 9A, nothing as yet outside Europe but still not bad for a telescopic aerial, actually fully retracted to reduce the dreadful RFI!

I have given special attention to the front-end circuitry of this design because one criticism frequently levelled at regenerative receivers is that they are too subject to blocking and

The RR9 Receiver – G0KJK

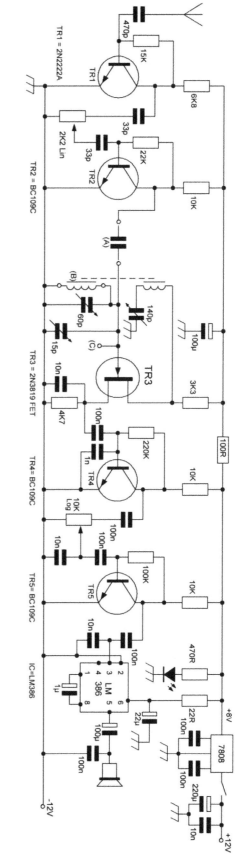
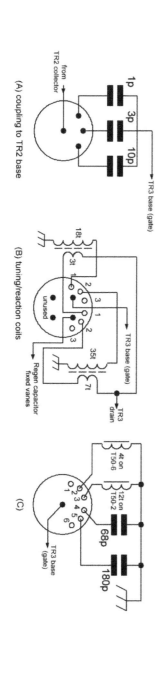

QRP SCRAPBOOK

overloading by strong signals to make them viable on our crowded amateur bands (although activity on these recently seems to have been low, conditions have not been good!). This receiver is not easily overloaded. It has three levels of basic aerial coupling, an active aerial principle before the RF gain control and variable resistor signal control. The AF gain control can be left quite well advanced and the strength of the desired signal governed by the RF gain knob. This latter will need to be kept well retarded when incoming signals are especially loud, typically during contests. Eight ohm loud speaker volume is such that the long-suffering XYL frequently exclaims – "Keith, can't you turn it down a bit?"! (or is my hearing not what it was?!!)

Please peruse the picture I give of the position of my own front panel controls. The receiver is built in an 8 by 6 by 4 inch aluminium box I acquired from Maplin Electronics. The block of four controls at left include the RF gain potentiometer and three rotary switches, one giving three coupling levels into the tuning coils, 1p 3p and 10p; another (4p 3w, only 2w used) choosing one of two tuning coils, 35t overlaid with 7t on a T-80-2 toroid and 18t overlaid with 3t on a T-50-2 toroid, which with a 60p tuning capacitor give the 40, 20 and 17m bands, and a third 1p 12w switch giving all the remaining six bands by the selection of different inductors and capacitors – including a 4t T-50-6 inductor for 6 metres, a 68p cap for 30m, approx. 180p for 80m and a 12t inductor on a T-50-2 toroid for 10, 12 and 15m. Experiment, experiment, experiment! Remember what you learnt for the RAE – that a capacitor across a capacitor lowers the frequency and an inductor across an inductor raises the frequency. Coverage possibilities are therefore endless! (If all this is as clear as mud, please blame my poor communication skills and send me an e-mail, to which I promise an attempt at an intelligent answer!).

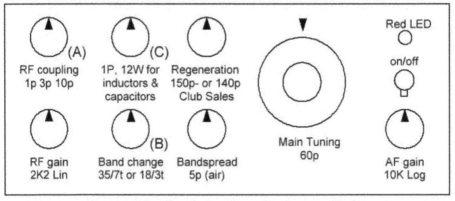

Rear panel has 3 sockets: antenna input, power 12v & speaker
Tip - Look out for airspaced variable capacitors at rallies!

PART 3 RECEIVERS

W1FB MEMORIAL ENTRY

The Chopping Board Receiver
Peter Parker, VK3YE. <parkerp@internode.on.net>

Features

- Tunes 80 and 40 metres, providing SSB, CW & AM reception
- All-discrete 3 transistor design with just 22 parts
- Sufficient output to drive low or high impedance phones
- Swinging link antenna coupling optimises sensitivity and selectivity
- No frequency pulling on strong signals or touchy regeneration
- Chopping board case forms front panel, carry handle and coil former

Circuit stages

- Selective front end with spiderweb coil
- VK4FUQ infinite impedance FET detector (MPF102)
- Audio amplifier (BC548)
- Colpitts variable frequency ceramic resonator local oscillator (BC548)

Construction notes

- Mark two concentric circles below handle of chopping board, approximately 6 and 8 cm diameter. With a jigsaw or hot soldering iron cut five slits between the two circles, equally spaced. Then drill holes for variable capacitors, sockets and antenna coupling winding.
- Wind detector coil with thin plastic covered insulated wire. Thread wire through slots, alternating in front and behind, like a basket weave, to form a spiral (see photos). Work from inside out, finishing after 13 or 14 turns (you'll probably remove some later during adjusting).
- Form antenna coupling loop about 7cm in diameter over the detector coil, using stiff but still bendable wire. Pass through holes and connect to antenna socket.
- Glue peaking capacitor near the centre of the board below its coil. Drill hole for tuning capacitor in piece of blank PC board material, glue, and solder centre connection (G) to it. Bridge outer terminals (A and O) to provide a 220pF maximum capacitance.
- Assemble the circuit board, dead-bug style, following diagram. Start with local oscillator. Then build detector and audio stages.
- Glue or screw circuit board to rear of chopping board. Make connections to remaining sockets.

QRP SCRAPBOOK

Testing and use

- With an HF receiver find the local oscillator's output. Check that it covers approximately 3.5 to 3.62 MHz when tuned. Vary 220 and 150 pF values if required.

- Connect outside antenna and tune peaking control slowly, listening for AM shortwave stations preferably at night. Reception proves the detector and audio stages are working.

- Listen for noise peaks near both ends of the peaking control's travel. This should be the detector picking up the local oscillator's fundamental on 3.5 MHz or harmonic on 7 MHz. Confirm with a local test signal and then try receiving CW/SSB amateurs.

- Can't tune 7 MHz? Remove turns from the detector coil. Or add them if 3.5 MHz isn't reached. When it's right you'll cover both bands; 11.8 turns on the prototype tunes 3.5 to 10 MHz.

- To use, peak the front end for the correct band, tune in stations with the local oscillator and re peak the front end if required. Moving the antenna coupling loop away from the detector coil sharpens the front end and helps avoid broadcast overload.

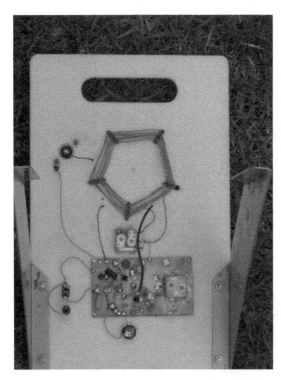

Underside of Chopping Board Receiver

Showing coil, circuit board and controls

PART 3 RECEIVERS

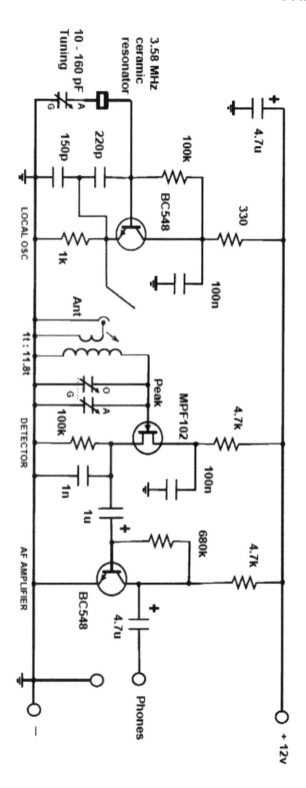

QRP SCRAPBOOK

Left:
Detail of Coil

Below;
The Circuit Board
(Ugly Construction)

PART 3 RECEIVERS

New High Performance Regenerative Receiver
Olivier Ernst, F5LVG, 2 rue de la Philanthropie, 59700 Marcq en Baroeul, France

I made this regenerative receiver for 5 amateur bands : 80 40 20 17 15m. With this Rx and a home-made transmitter, I made several SSB QSO between North America and France. It is possible to listen, without noticeable detuning, SSB QSO on 15m during 15 minutes. There is no hand effect, no common hum and no mains hum.

I will focus on the following points:

1 RF attenuator, mandatory for all regenerative receiver.
2 Very small coupling capacitor between the antenna, the RF transistor and the tuned circuit to avoid overloading (I use my transmitting antenna).
3 Plug-in coils without coil forms to obtain very high Q. The coils are easy to make : only one coil without tap for each band. I use 4 pins DIN connectors.
4 High C "oscillator" with NP0 capacitors to obtain a high frequency stability.
5 Band spreading with small capacitors in series with the variable capacitor.
6 High capacitor value between the tuned circuit and the detector to avoid mains hum.
7 Very short connection for the dot and dash lines.
8 Fine tuning with a 1N4007 diode.
9 1N4148 : transistor protection during transmission.

Diameter for the coils : 22mm

80m, L=11 turns, Ct=470pF, Cp=122pF
40m, L=5 turns, Ct=552pF, Cp=55pF
20m, L=3 turns, Ct=320pF, Cp=25pF
17m, L=2 turns, Ct=440pF, Cp=16pF
15m, L=2 turns, Ct=253pF, Cp=16pF

Wire for 80 40m : diameter 0.5mm.
Wire for 20 17 and 15m : 2.5mm2 installation cable.

The picture on the front page shows the layout. The size is 15x20 cm. I used 10M ohm resistors for the connection points isolated from ground !!! (stand-offs) 10M ohm is nearly infinite....

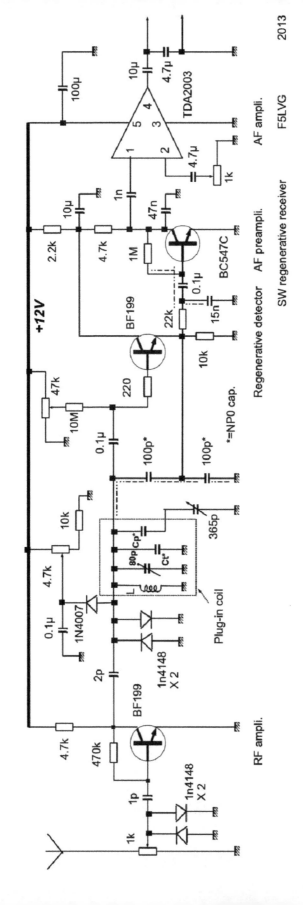

PART 3 RECEIVERS

40m SSB/CW Receiver.
Andy Choraffa, G3PKW, 1 Windsor Rd. Roby, Liverpool. L36 4NG

I decided to build a small pocket rx which I could take on holiday. Light weight and small for air flight and cover 40m and maybe 80m later. I wanted it to be as economical as possible with a small 9v battery. I decided to build a circuit similar to the GQRP club 'Sudden' Rx. I wanted it to be easy to reproduce with common available parts and on Vero-board. I didn't want to start designing a PCB with all the etching procedure.
Vero board proved to be quite adequate and no track leakage problems were evident. Initially I changed a few component values from the original design to optimise performance.
Component values for both the 602 Gilbert cell balanced mixer and the 386 AF amp were decided on from my experience and calculations around the original data. These are slightly different from the original GQRP club design. The oscillator circuit in the SA602 was designed for PMR use with a xtal and works well. The design impedances around the SA602 are low as is evident in the 'Sudden' rx. Note the 2 uH as used for the osc coil design, yet 6 uH for the input coil. Both these tuned circuits are operating at 7 megs. I built a VFO around the SA602 but found it was not very reliable in starting. I think this was due to these low impedances which the chip was originally designed around. As I experimented with a xtal which worked well. I could have continued to wrestle building an osc using the SA602. But its lack of purity encouraged me to build a separate FET osc.
The spectral purity of the original osc design I found to be poor The 3rd harmonic at 21 megs was only about 12 dB below the 7 meg wanted. It also had some second harmonic at 14meg., about 25 dB down. In PMR use with a xtal this wouldn't be evident or a problem.
As the xtal osc in those applications is typically an overtone variety. I built a separate FET osc and it started very well with a good clean spectrum. It was very clean with all harmonics at least forty dB down. The FET only needed 1 and half mA at about 4v to operate without any problems. I stabilised the osc supply with a couple of 4148's plus the red Indicator LED. The 2k2 resistor was the highest value to maintain enough LED brightness to be visible. To make the LED brighter a lower value will be needed.
The 4148's were in plentiful supply and cheap, but I could have used a single 3v zener in its place. The 602 only needs about 6v at two and a half mA so was fed via a 270 ohm and suitably decoupled. The 386 draws a quiescent three and a half mA so the total drain from the PP3 battery is about 8 mA. The picture of the original prototype shows the rats nest construction after many mods. The FET osc was built suspended around the osc coil assembly as shown The Aladdin former had the thirteen turns held firm by small pieces of heat shrink sleeve offcuts. Then the three turn link was loosely wound over the top.
The 15 puff ceramic from the 30 puff air spaced variable gave nice coverage of 40m over the full 180 degrees. This enabled tuning CW / SSB without reduction drive, possible, and just a large red knob sufficed.

*Note the oscillator coil could probably be replaced by a Spectrum 5u3L coil.

QRP SCRAPBOOK

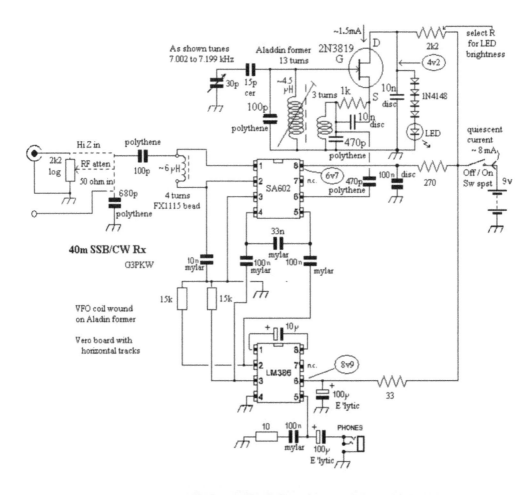

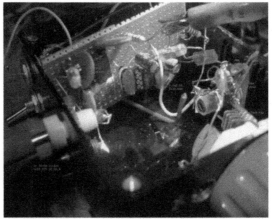

Inside the receiver.

PART 3 RECEIVERS

A Valve Regenerative Receiver
Barry Zaruki M0DGQ, 26 Heathfield Rd. Kings Heath, BIRMINGHAM. B14 7DB

This simple receiver uses three Russian rod pentodes, a marvellous device from the cold war era and in plentiful supply to this day (eBay). The set covers 1.6MHz to 8MHz, 730kHz - 1200kHz BC and 472kHz using plug in coils for the band in use.
With a reasonable aerial the set is very sensitive and drives a loudspeaker at good volume, roughly 300mW. Bandspread tuning is also used to enable easy tuning of amateur stations.

Circuit Description

The aerial is connected directly to V1 control grid, no grid tuned circuit is used. V1 is reflexed, it amplifies RF and AF simultaneously, the
gain of this stage is controlled by its G2 voltage (VR1). The main purpose of V1 is to prevent oscillation from the regen stage being radiated by the aerial causing interference to other stations.

At RF the anode load for V1 is tuned circuit L1 - VC1, while at AF the anode load is R3. Amplified RF from V1 is coupled to V2 control grid. V2 is configured as a regeneration stage, degree of regen is control by G2 voltage via VR2.

Demodulated audio is taken from V2 anode via C14. R11, C2 form a low pass filter thus eliminating any RF from V2 anode passing to V1. Amplified audio is taken from V1 anode via C6, R5 - C7 form another low pass filter to remove any RF/oscillation being passed to V3. A 1J29b is used as an audio output stage providing a few hundred milliwatts of output power. V3 anode load is a small 100V line transformer, the 250mW tap is used as this gives the best match and loudest volume, it presents a load impedance of roughly 40K. Standard output transformers work but with slightly reduced volume.

A range of plug in coils are used for L1, these are constructed from a piece of broom stale with copper wire pins on the bottom arranged in a B9A
pattern. 1mm holes were drilled in B9A formation and suitable stout copper wire was pushed and glued into the holes, I used a two part epoxy resin.
The wire used for the coil winding is 32 swg enamelled copper wire. On the broadcast band coil a "cheat " is included, a switched 470pF capacitor is used to receive CW transmissions on the new 472kHz band. The switch and capacitor are mounted on the coil (see photos).
 All coil winding details are given in the circuit diagram, the frequency ranges are approximate so the odd turn may be added / subtracted depending on stray capacitance etc.

The set works well with a dipole (at home I use it with a 80m dipole), end fed long wires will also work fine provided a step down RF transformer is used in line with the antenna, otherwise all that will be heard is loud mains hum due to the first stage acting as a audio

QRP SCRAPBOOK

pre amplifier with a long wire connected to its input. Alternatively a 2.2mH choke connected between the antenna terminal and ground will suffice.

In use

The set is easy to use and its performance is very good indeed for such a simple set. Amateur stations are easily copied, both ssb and cw. Broadcast stations romp through, VR1 is useful here as it controls both RF and AF gain.

Note
The suppressor grid (G3) is internally tied to the filament/cathode in the 1J18b tube, this is not shown on the circuit. If you are using a battery to power the tube filaments do not use a 1.5V battery, use a NiCad (1.2V).

PART 3 RECEIVERS

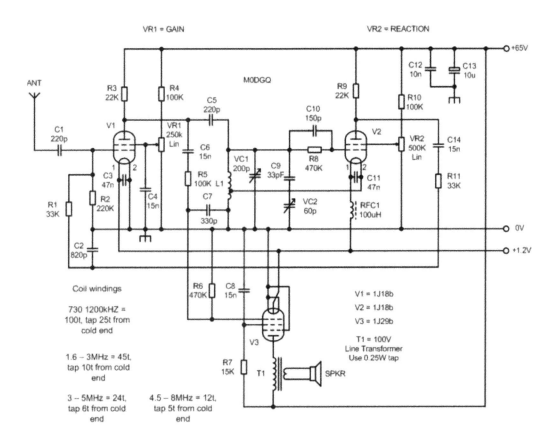

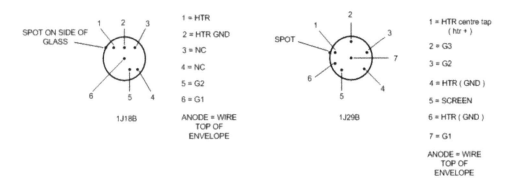

QRP SCRAPBOOK

A Primer for Software Defined Radio (SDR) using the RTL2832U R820T Dongle
Ken Marshall G4IIB <ken.marshall1@btinternet.com>

The DVB-T TV dongle has been around for several years now. These are very cheap (typically under £6 from eBay) other outlets are available)) and can easily be converted to a very versatile general coverage receiver covering from 0 to 1.7GHz. A useful and interesting addition to any shack as, used with the right software provides a selection of receiver modes and useful filters. The DVB-T dongle discussed in this primer is one utilising the RTL2832U and R820T chip set. The RTL2832U is an analogue to digital converter and the R820T is a tuner chip covering 24MHz to 1.7GHz.

This primer is intended to guide you through the purchase of a suitable dongle and the setting up of suitable free software. In addition I will also describe a simple hardware modification and software parameter change to get the dongle to work on the lower HF frequencies.

Hardware required:
    An RTL SDR Dongle
    A computer running windows XP/vista/Win 7/8 with dual core CPU, USB 2 or higher and minimum 1Gig of memory Note PC must be running Microsoft.Net (but probably already is) and 7ZIP.
    A suitable antenna and ATU

You require a DVB-T TV dongle, make sure that it has the RTL2832U and R620T chip set before you purchase. The cheapest place to get one of these is eBay. Here one can be purchased for as little as £6 if you shop around and are prepared to wait for one shipped from China (worth the wait in my opinion). However, for the more cautious or impatient buyer a few UK outlets are selling these but you will usually pay a premium for UK sourced dongles. I have also seen these at rallies for around the £10-15 mark, well worth consideration. Opposite, for reference, is a picture of what you need to purchase.

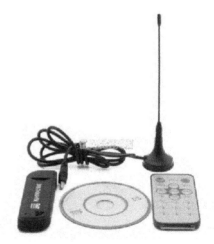

Note the MCX antenna connector; this is desirable as it provides better loss figures at the higher frequencies, although it is likely that you will change the antenna socket if you decide to put it in a metal box.

Once you have the dongle we can get it up and running using one of the FREE SDR software products such as HDSDR, SDR#, GQRX and others. This primer will concentrate on the HDSDR software as it provides both RF and Audio FFT *(Fast Fourier*

PART 3 RECEIVERS

Transform) waterfalls and a range of sophisticated filters. However, all of the above software products have their plus points.

First of all unpack your dongle. THIS BIT IS REALLY IMPORTANT pick out the accompanying CD and carefully remove this from the packaging now THROW THE CD IN THE BIN. The CD is useless for SDR and even if you intend using it to receive TV, up to date software for this is available for download on the web. The other item we can get rid of is the remote control, whilst this will work with the TV dongle it will not work as an SDR radio, put it in your "might come in handy bin". The antenna can still come in useful.

Software Installation

Download ZADIG driver software from http://zadig.akeo.ie/ you will need 7zip to extract it - note different versions for windows version i.e. Win XP. Plug in your RTL2832U stick. Follow this User Guide https://github.com/pbatard/libwdi/wiki/Zadig. Basically list all devices in the option drop down menu to see your stick. This will usually be a bulk-in, interface (interface 0) device or in some cases simply rtl2832u. Once selected proceed to install the driver. Don't worry about the *Windows can't verify publisher warnings* just install it anyway. If you have more than one dongle you may have to rerun zadig on another USB port. I could not get two dongles to run on the same port, but I have recently come across some that do. A case of try it and see.

Note: the following steps will not have to be redone for the second dongle. You should only need do these steps once.

Download HDSDR from http://www.hdsdr.de/ download button is at the bottom of the page.
Install HDSDR follow the instructions from this site
https://sites.google.com/site/g4zfqradio/installing-and-using-hdsdr

Download ExtIO_RTL2832U.dll file from
https://github.com/josemariaaraujo/ExtIO_RTL press download button extract then open the release folder you have a choice of 32 or 64 bit versions. Copy the appropriate version into the HDSDR directory i.e. C: Program Files (x86)/HDSDR

Plug in your dongle and start HDSDR. There is no official user guide but the previous instructions at https://sites.google.com/site/g4zfqradio/installing-and-using-hdsdr will help in setting up the software parameters. In addition the HDSDR FAQ is useful. If you hover your cursor on any HDSDR buttons a brief description/instructions on use will appear. You will find that HDSDR is quite intuitive to use.

Once you plug in a suitable aerial and start HDSDR your dongle should now be able to tune from 24MHz to 1.7GHz and you can listen in AM, ECSS, FM, LSB, USB, CW and DRM modes. Prior to it reaching operating temperature it does drift a little but the dongle becomes quite stable after about 4-5 minutes of actual receiving i.e. not just plugged in on standby.

QRP SCRAPBOOK

Getting the Dongle to work on lower HF frequencies

Once you prise open the dongle you should be able to identify the following.

The RTL2832 analogue to digital converter can tune from 0 to 29 MHz. However, for TV reception purposes its tuning range is extended from 24 MHz to 1.7 GHz as it is fed by the R820T tuner chip, the lowest frequency that this chip can tune being 24Mhz. SDR software has the facility to allow the user to change the sampling mode. Using the tuner chip the SDR software (in this case HDSDR) uses Quadrature sampling. In the previous steps above you may recall that we placed ExtIO_RTL2832.DLL in the HDSDR directory. With HDSDR running you will see that the ExtIO tab is highlighted bright blue (next to Tune digits). If you left click this tab a panel will open up. Under the second heading "Direct Sampling" you will see that it is Disabled this gives us Quadrature sampling as we are using R820T. The other method of sampling is Direct sampling, in this method the R820T tuner is bypassed and the RTL2832U looks for a signal on its input pins 1&2 (I channel) or 4&5 (Q channel). Again with HDSDR running, but STOPPED, and clicking on the ExtIO tab you can see that we can enable Direct sampling on either the I or Q inputs.

The R820 feeds pins 1&2 via capacitors C33 & C34. Pins 4&5 are not used i.e. there is no connection. If we connect an aerial to these pins then we can receive the lower frequencies by changing to the direct sampling method.

Note: I use the term Quadrature sampling purely from a software parameter setting point of view. In this dongle actual physical Quadrature sampling is not used as only one channel, the I channel, is connected. Dongles using the E4000 tuner chip (and others) do offer full Quadrature sampling as they feed signals to both the I and Q channels of RTL2832U. E4000 dongles have become obsolete but can still be obtained but are much more expensive than the R820T variety and cover slightly different frequencies 50MHz to 2GHz. Furthermore, from an HF mod perspective using one of these more expensive E4000 dongles would be like using a broken pencil. POINTLESS.

How good are your soldering skills?
I thought that mine were reasonable. They were rubbish! In theory you could use the unused pins 4&5 and solder a toroid to these and connect a separate aerial socket to the other end of the toroid, this way the dongle would tune 24 MHz to 1.7GHz using

PART 3 RECEIVERS

Quadrature Sampling and 0 to 24 MHz using direct sampling. If you can solder on to these pins, then good on you. For me that was impossible, so I opted to solder my toroid (8 turns triflar wound on FT37-43) on the pin ends of the two capacitors C33 & 34. The reason for using both pins on the RTL2832U was that it creates a differential input of the + and − portions of the signal. In attaching the toroid I accidentally bridged these capacitors with a blob of solder. In removing this blob I also succeeded in removing both capacitors (looks like I need an even smaller bit on my iron). Ah well, I was going to dedicate this dongle to receiving HF only anyway. To retrieve the situation I managed to solder a single piece of wire to the pin end of where C34 had been i.e. on the track leading to pin 1. The toroid will have to wait until I upgrade my iron and soldering skills. I soldered the other end to the static protection diode D8 on its contact next to C13, thus the aerial socket was now connected directly via D8 to pin 1 of the AD converter. Connecting the aerial socket to my ATU and using the direct sampling method on I input, the dongle now tunes 0 to 24 MHz.

Impressed with the results of my botched HF mod a fellow amateur friend asked if I would convert a dongle for him. A good chance to practice my soldering skills. After purchasing a temperature controlled iron with a finer point, and with the advice offered by G8RIW in Sprat 161 regarding the use of blue tack to hold the wires in place before soldering, and with the aid of two pairs of glasses and a magnifying glass (it's awful getting old) I have managed to solder a toroid to C33 & 34. For the toroid I used an FT37-43 with 8 turns triflar wound of 33 SWG magnet wire (I found this to be better than 5 turns). For connections follow the instructions in the image shown below.

Both of the above HF mods work but the toroid mod performs slightly better, I believe due to the use of the differential signal. See the results of my efforts below. And you can get the lid back on; however, they are best housed in a metal box.

I also have a dongle with an up converter but the above direct sampling mods perform better.

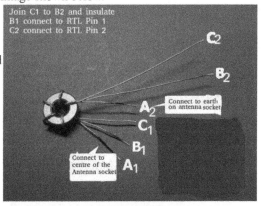

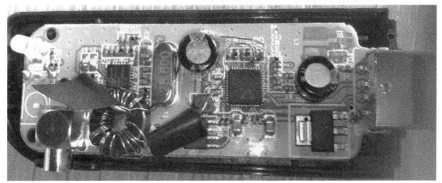

QRP SCRAPBOOK

Improvements

Further improvements can be achieved by adding a preamp for HF reception and constructing band pass filters for your favourite bands, a good site for calculating these is http://www1.sphere.ne.jp/i-lab/ilab/tool/BPF_C_e.htm

The T820 does contain a preamp but it is noisy (typically 8db) this can be improved especially for reception at the higher VHF frequencies by adding an LNA ideally at the antenna end. If you are not using the dongle for TV then I suggest removing the IR sensor (3 pinned black thing next to aerial socket) as it is a source of noise especially at the higher frequencies. The dongles are best screened by being mounted in a suitably robust metal box. They run hot this is normal.

Station Monitor and Panadapter

These cheap dongles can be a useful second/umpteenth receiver for the shack. They are also useful for monitoring your signal via both the RF and audio spectrum displays.

In addition, if your rig has an IF output, connecting this to the dongle aerial socket means that the dongle can be used as a panadapter by tuning HDSDR to your first IF frequency, then anything you tune on your rigs receiver will appear on the RF FFT display, allowing you to see a panoramic display of the band.

The RTL2832U ADC is only an 8 bit device in the future 16 bit dongles are likely to a appear on the market these will offer an improved performance. If the price is right watch this space.

Have fun with your dongle!

Note: if you want to use a "sound card type Receiver" like the one in SPRAT161 you only need install HDSDR and plug the RX in to the sound card. Follow the user guide https://sites.google.com/site/g4zfqradio/installing-and-using-hdsdr. You only need the zadig driver and the DLL if you are using a dongle. A good sound card helps i.e. one with a bandwidth of circa 50KHz or better still 100KHz.

Screen shot of dongle after HF conversion receiving JT65-HF on 40M

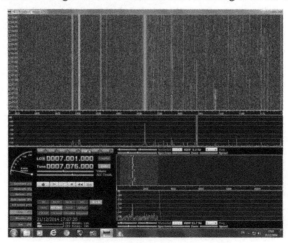

PART 3 RECEIVERS

More on using the RTL2832U R820T Dongle
Tony G4WIF <g4wif@gqrp.co.uk>

Elsewhere in this Sprat, Ken G4IIB shows how to convert this very prolific device often seen at rallies. I got mine for £15 a year ago and just recently, our USA DX rep, Dave W7AQK ordered one for $9. So we aren't talking about a big investment.

Having followed Ken's instructions I was amazed at how well it worked on HF considering the simplicity of the conversion and I started to wonder what else we could use it for. I am blessed in having a spectrum analyser but this tool won't feature in every shack - and just being able to see at a glance if a filter is performing at all is just plain useful.

So I injected wideband noise into a 40m band pass filter and looked to see how the output appeared on the TV dongle SDR display. Below you can see the results.

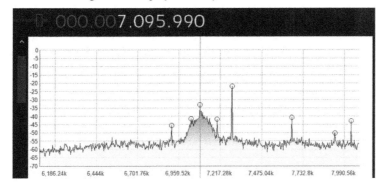

(snapshot taken using "SDR Sharp" software as an alternative to HDSDR used by Ken)

Now I'm not at all suggesting that this is a precision piece of measuring equipment or that the SDR software dB markings should be taken as anything more than guidance. This is what you could call "indicating equipment" and simple test gear is a lot better than none at all. On similar lines is the noise generator that I used. It is the only one I've ever needed and I built it quite a long time ago from a design by Tom N0SS.

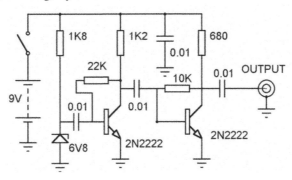

To save space in Sprat from unnecessary duplication I have put some notes and photos of my version of Ken G4IIB's project on the Sprat pages on the club website.

SPRING 2015 SPRAT 162

QRP SCRAPBOOK

That RTL2832U R820T dongle again.
Tony G4WIF

In the Spring 2015 Sprat Ken G4IIB started this TV dongle madness with a brilliant article showing a simple mod which allowed a cheap USB TV dongle to work on the HF bands.

Without modification the performance fell off below 24MHz. However with the modification, and the use of one of the free software defined radio applications, we can have an all bands radio covering from medium wave up to around 1.7 GHz.

In the same Sprat I wrote how you could use this setup with a homebrew noise generator to get a feel for whether band pass filters etc were working or not. The photo to the right shows just that using the "SDR Sharp" receiving software.

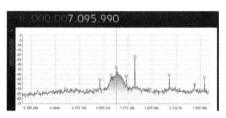

I began to investigate what other test software might be around to use with the dongle and found an excellent free application called "RTLSDR Scanner". The only problem being that I couldn't find a way to use it at HF frequencies using Ken's modification.

With a birthday coming up I treated myself to an HF to VHF converter which is sold specifically for use with these TV dongles. It is called the "Ham it up convertor" from a company called "NooElec". You can buy it through Amazon or EBay for around £35 and it is worth every penny because now HF performance with the dongle is very impressive. For a total outlay of £45 (including the dongle) you can now have a great software defined radio receiver for a price that we wouldn't have thought possible a few years ago.

Using my wide band noise generator I injected a signal into a band pass filter for 40 Metres and scanned it on the new dongle/NooElec converter combination with the RTLSDR Scanner software. This is the test result - which is a great deal improved in terms of information than my original test shown in the Spring Sprat 2015.

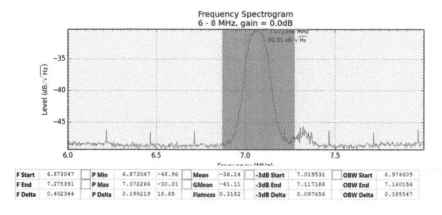

PART 3 RECEIVERS

Now as I wrote previously, I am not claiming that we have a spectrum analyser with tracking generator here and the accuracy of the dB markings may not be massively accurate. However, if you wanted to build a filter from one of the many published designs, and needed to check that it was actually working where it was supposed to, you would get a pretty good indication plus some idea of the bandwidth. Much better at least than you would if you had no test equipment.

The NooElec converter is well designed and has some good filtering built in for HF reception, but for filter testing using "RTLSDR Scanner" perhaps something simpler, based on an NE602 would serve?

Another experiment I tried was to inject a 1 MHz square wave into the setup and compare it to my spectrum analyser using 40dB of attenuation (in both tests) so as not to overload the converter front end.

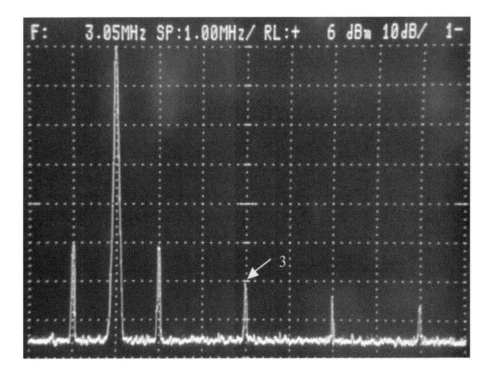

[Spectrum Analyser Snapshot]

QRP SCRAPBOOK

At the centre frequency of the left hand photo the harmonic at 3 MHz is about 10dB down on the fundamental (each division being 10dB).

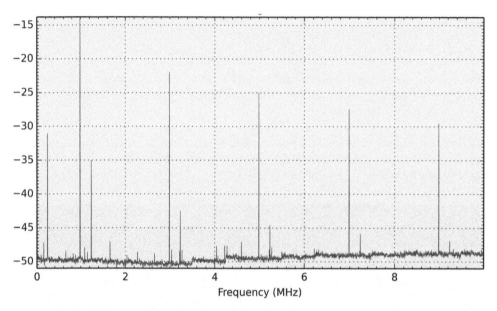

[RTLSDR Scanner Snapshot].

In the RTLSDR Scanner photo you can see that the 3 MHz harmonic is about 9dB down. There is a delightful American expression, "that's close enough for government work".

You will observe that with RTLSDR Scanner you can get some artefacts such as those shown at 0.15 MHz above each harmonic. I think we can make some allowances for inexpensive hardware and for software that is completely free. Some of the "sproggies" I've seen are due to radiated interference from the computers own screen and I've had to turn it off while RTLSDR Scanner is sweeping.

With care this setup can show us a little more information about what is going on in our radios and at a great price.

Finally this setup will also allow reception of all kinds of data modes and I have placed a longer description of how to achieve that on the "Sprat" page on the club website. I will also show some more screenshots of some filters I've tested there.

Web links: eartoearoak.com/software/rtlsdr-scanner
www.nooelec.com and www.gqrp.com

PART 3 RECEIVERS

The Tin Can 2
A Two Transistor 80m Receiver by John Hale VK2ASU
Reproduced with thanks from Lo-Key No:102 (VK QRP Club)

The idea for this little receiver as a two active device contender had been in my notepad for quite a while. During a week of rain I was stripping a couple circuit boards when I found an MPSA 14 Darlington transistor. The notes came out and a two transistor direct conversion receiver for 80 metres using mainly junk box parts was built. Connected to my inverted V, it works so well that I figured it was worth sharing.

I built the circuit onto a piece of double-sided circuit board about 1 x 2 inches (2.5 x 5 cm) I cut one side with a hacksaw to form many copper islands – chess board style. Because the circuit board was double-sided the uncut side was earthed, and any perimeter island could then be earthed by soldering a piece of wire to it and the underside of the board. The components were soldered to the copper islands following the circuit schematic. The antenna input coil was made from an oscillator coil taken from a discarded transistor radio which also provided the polyvaricon tuning capacitor. The oscillator coil usually has a red adjustable ferrite core. This was carefully taken apart and the 100 turns of fine wire wound onto a pencil for rewinding the new coil. If this idea is followed take care not to kink the wire. The new coil was wound with 4 turns for the antenna input and its ends connected to the side of the coil base with 2 lugs. The secondary winding was in the same direction over the top of the first winding, and started at the earth lug (on the 3 lug side of the coil base). After winding on 10 turns a tap was formed by connecting to the centre lug and after another 16 turns the coil terminated on the remaining lug. I have made a few 80 metre antenna coils using these old oscillator coils and using the same number of turns. They have all worked well, and simply adjusted with the ferrite slug. If you try to build this receiver and coil doesn't resonate, increase or decrease the value of C3. If you have a signal generator and a CRO this is a snack, otherwise trial and error will eventually get you there.

The Hartley oscillator coil was wound with 50 turns of wire rescued from the yoke of a TV picture tube,. The winding filled the Amidon T50-2 toroid. The tap is at 12 turns. The reduction drive for tuning was made from an old long shaft potentiometer. The back was taken off and the middle pin removed so that the shaft rotated freely. The dial cord and tuning drum were retrieved with the polyaricon capacitor from the discarded (bedroom clock) radio. Unfortunately the tuning knob spindle was moulded into the case and was not recoverable. The dial cord was wound around the pot shaft 3 times and worked quite smoothly, to give a 6:1 reduction.

QRP SCRAPBOOK

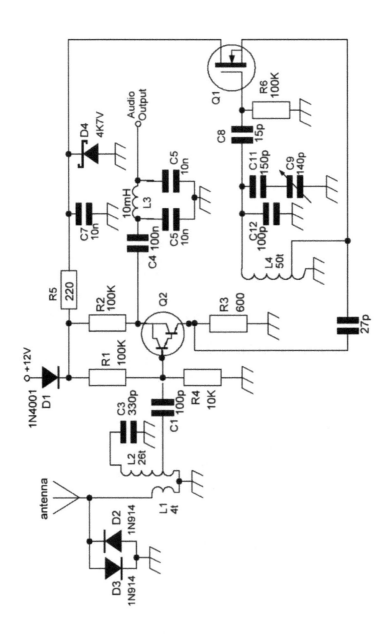

PART 3 RECEIVERS

The receiver worked great with a crystal earpiece, but I later decided to make an audio amplifier to drive bud style headphones. The headphone amplifier was built onto a piece of double-sided PCB about 1 x 1 inch (2.5 x 2.5cm) in the same way as the receiver board. Although BC547 transistors are shown, just about any NPN transistors will work in this circuit. A pair of bud style headphones was connected in series to give a 64 ohm impedance. I cut the stereo plug off and put a 3.5 mm mono plug on. The series connection was inside the plug and insulated to prevent shorts. Thin coax was used between the receiver and the amplifier.

Alternatives to the T50-2 toroidal coil and the polyaricon capacitor can be used. For the coil, get a drinking straw and wind on 75 turns of 26 gauge enamelled copper wire tapped at 19 turns. Then arrange a brass screw to screw in and out of the coil. Be careful soldering though, as it is easy to melt the straw. Remove capacitors C9 and C11 and replace them with about a 10 – 90 pF trimmer capacitor. Adjust to obtain oscillation at 3.5 MHz when the brass screw is wound out. As the brass screw is moved into the coil the frequency will rise – the opposite to the effect of a ferrite core. A bonus of this system is that further tuning reduction is not needed.

Alternatives for the Darlington are MLPSA12 and MPSA13, both of which I have tried and work fine. There would be many others. The MLPF102 could be substituted the a 2N3829 or a J310. Just try what you have. The 1N4168 diodes could be any small signal silicon diodes including 1N4148's. The 4.7 volt Zener would be substituted with a 5.1 or 5.6 volt one – I just found 4.7 volt first. Another option would be to add another active device, a 78LO05, and forget the two active device concept. The 10 mH RFC was one I had, but you could try a 1 or 2 mH choke. It is there with the bypass capacitors to stop RF going to the earpiece or audio amplifier and to shape the audio a little. As an afterthought D4 was added to ensure against damage from a wrong polarity connection.

My QTH is in Parramatta and only a few miles away are powerful broadcast band transmitters which often cause breakthrough in simple receivers. To my surprise there is no sign of broadcast breakthrough in the little receiver. This probably due to the amount of RF injected at the emitter, as well as the diodes at the antenna input reducing the maximum RF level to the circuit. The diodes have no effect on normal strength 80 metre signals.

I perceive the MPSA14 to be working with the first transistor of the Darlington pair working as an RF amplifier, while the second of the pair is mixing and performing an audio amplification function. This little receiver has given me much listening pleasure at nights in the shack while working on other projects. I hope there are others who will give a version of this receiver a go. Please email to me at vk2asu@wia.lorg.au for any help or to make comments.

Simple 80m Modular Receiver
Peter Howard G4UMB 33 West Bradford Rd. Clitheroe Lancs

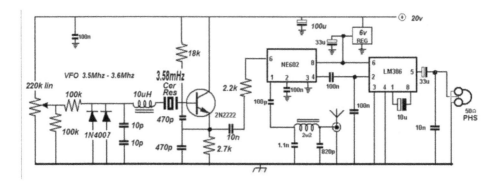

Here is a simple direct conversion receiver for tuning from 3.5 – 3.6 Mhz covering the CW section of the 80m band. In order to get that range of tuning I have used a 3.58 Mhz ceramic resonator with a 10 micro Henry coil in series with two 1N4007 rectifier diodes in parallel to act as varicaps. You must use a 20v supply to get the full frequency range.

Having built this VFO (VXO) part first I then made the other stages of the receiver on separate plug-in boards so that each piece could be tested separately and removed from the small box to work on them in case a fault develops later. The circuits are built on strip board and the plugs and sockets are made from turned pin IC sockets.

The circuit is based on the receiver in a FOXX trx I had already built which has proved successful but lacked a VFO. It works best with an ATU because as you can see the tuned input circuit is far from ideal and this will greatly improve its performance.

PART 3 RECEIVERS

Simple DC RX covers 8 bands 40m to 10m with 27 components
Andrew Davies G6JLH

This DC RX breaks with the conventional approach to direct conversion receiver design and covers more with less. The component count is down to just over 3 components per band!

The VFO covers 7 to 7.2 MHZ this makes the harmonic mixer sensitive to 40m on the fundamental frequency and 20m, 15m, 10m on harmonics. With a suitable band setting variable capacitor the 7Mhz VFO can be reduced to 5.05MHZ to cover 10.1Mhz for 30m coverage on the second harmonic. Likewise the third harmonic of 6.022 MHz could be used for 17m and fourth harmonic at 6.225Mhz for 12m.

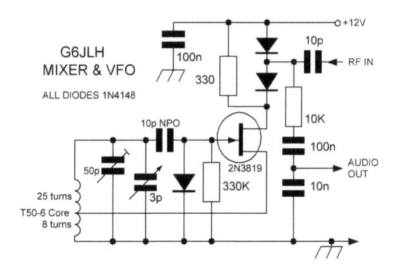

The RF gain pot is provided for volume control and protection against overloading and AM breakthrough. In strong signal conditions without the RF gain control you have an Atmospheric radio. This gives a new mode of operation and another band.

RF Amp Pre-Selector
I used only a single tuned circuit for simplicity of construction. I found that pushing one toroid inductor to its limit will tune 40m to 10m without the

need for switching. The RX front panel needs the ham bands marked out. The alternative idea of using a rotary switch and a bank of trimming capacitors would increase complexity. Some experimentation may be required for highest stable gain regarding the inductor winding phasing. Try to keep the pre-selector away from the VFO or use some Faraday shielding. The 2N3819 FET used common drain giving lower gain but it is more stable.

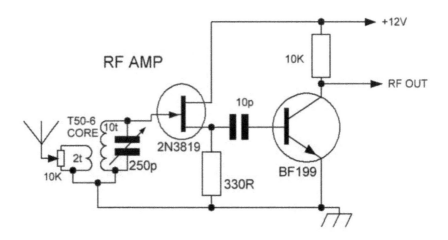

The next stage is a BF199 common emitter RF transistor. This RF amp more than makes good the losses of the mixer when operating on harmonics.

The amplified RF is few to the mixer I have not seen this configuration before so this could make it the first DC RX with a Davies mixer. The audio output is fed to a BC549C low noise audio transistor. The crystal earpiece provides sensitivity without being dangerously loud avoiding the need for an AGC circuit.

*** Sorry to say Andrew ... I think you have been beaten on the mixer. I recall versions of this circuit appearing on a website. I cannot quote chapter and verse but suspect others will.***

I used a 2N3819 FET for the VFO because I have a lot of them but other types can be substituted.

The 3pf tuning variable capacitor is not a common junk box part you may experiment by using a larger value connected between the FET source and ground on the tuned circuit tap.

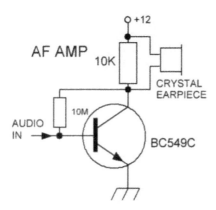

New Mixer Design (?)
This mixer came about as a result of trying to build a simple superhet receiver with a single FET unbalanced mixer. The considerable unsuppressed output from the VFO caused the crystal filter to ring and over loading of the IF amplifier. After some thought I came up with a simple diode mixer that fixed the problem. The next step was to test the mixer's suitability for use in a direct conversion receiver by testing its ability to resist demodulation of amplitude modulated broadcast band signals. A capacitor of 0.01 uf was added to the output taking RF to ground so only AF is present at the output.

Circuit Description
The two diodes in series with the FET drain act as an RF switch. When no current is flowing the diodes offer a low impedance to ground. The result is the audio frequency output.
Because it works on switch mode it is also sensitive to harmonics at reduced efficiency.
Even though the diodes are driven by a sine wave not a square wave it took 100mV AM signal before the audio signal could be heard breaking through. So the diodes will not react as an envelope detector easily. This made the mixer suitable for more testing in a DC RX.

QRP SCRAPBOOK

CS5 Direct Conversion Multi-band SW Receiver
Keith Ranger G0KJK

Poor HF conditions and low activity on the bands made me think of putting together a low-cost, no-frills, easy to build SW receiver that would pull in worldwide DX and cheer me up – as I reluctantly gave my QRP transmitter a rest. The result has been the circuit featured here. Carefully built and properly operated, it should be found an effective simple amateur bands receiver, with good SSB and CW resolution and capable of covering a number of amateur bands.

I have called it the "CS5" because "CS" stands for "Club Sales" and it uses five crystals or ceramic resonators all available from Graham G3MFJ, who faithfully serves us all with these vital components.

The five transistors used in the CS5 circuit could be 2N3904 of 2SC536 devices available from Graham at a mere 50p for each lot, and the LM386 IC would add a mere 45p to your bill. In other words, you get all of them for less than half the price of a tea-shop cup of coffee (apart from P and P of course)! Who says amateur radio equipment has to cost the earth? Actually, as the CS5 circuit diagram reveals, I used BC109C and 2N2222A devices because I had them available, but either the 2N3904 or 2SC536 should do the job well, just peruse the performance specifics given by Graham on the back outside cover of your SPRAT.

The polycon variable capacitors sold by Graham, a 285pf and a 140pf example, will ideally provide the frequency agility respectively of both bandpass filter and local oscillator, as per the circuit diagram. You will need to find a top quality of 5pF air-spaced variable capacitor for the bandspread capacitor in the Local Oscillator, to give easy resolution of SSB signals especially. Check the Web and radio mags or visit the Rallies!

Now for some circuit details, etc. I used a desk-based 3ft long telescopic aerial, but your creativeness may suggest better mounting alternatives to my aluminium right-angled flange, which is secured to an off-cut wooden base by two wood screws. A hole large enough to hold the telescopic aerial's knuckle using grommets completes this assembly.

Such a short aerial has both advantages and disadvantages. It does not provide a strong input signal but it does prevent overload and fits the challenge of our G QRP Club ethos – "It is vain to do with more what can be done with less". An indoor aerial is also convenient in space-using terms. To make sure the CS5 RX has real sensitivity that does not over-rely on its AF stages, where such dependence could produce instability and/or microphony, a powerful two stage RF amplifier precedes the Mixer.

PART 3 RECEIVERS

The overall effect is very pleasing. AM breakthrough does occur from time to time but usually never enough to cause reception trouble. The gain distribution results in a real DX-puller receiver.

The Mixer in this circuit avoids the assembly of trifilar coil windings and matched diodes, which some Club members have described as something of a headache and easy to get wrong. The 2N2222A (or 2N3904 or SC536) is ideally suited to Mixer use, the RF stages and Bandpass Filter feed into the Mixer's base and Local Oscillator connects to its emitter, please consult the circuit diagram. An added advantage of this arrangement is that it provides some overall amplification rather than an insertion loss.

I am a very amateur radio amateur with no formal electronic training, so in circuit design and construction all that interests me is – "Does it work?" and how can I very conveniently put the whole show together? If you hesitate to go down the printed circuit or SMD route, why not use my simple method of using tag boards? Always gracious and helpful Will Outram, of Bowood Electronics, who advertises in SPRAT, sells superb tag boards. You will see examples of their use in the photo of the inside of the CS5 contained in this article.

Possibly the greatest difficulty, if circuit assembly gives you such, will be encountered when you put together all the features of the Local Oscillator stage, which will require 5 switchable crystals or ceramic resonators, all available from Club Sales, a series of small axial inductors, also from Club Sales, two 1P 12W rotary switches and two variable capacitors. Please consult the circuit diagram.

The five crystals or ceramic resonators are 3580kHz and 14320kHz resonators and 5262, 7030 and 1011kHz crystals. The wide-ranging bandpass filter, using inductance or capacitance switching to extend its tuning range, will tune to the fundamental frequencies or harmonics of all these five devices. This means many amateur bands are in part or in entirely coverable. As an example the 3580kHz resonator will tune up to 3620kHz in this circuit. The fifth harmonic is found in the 17m band, and reception at that frequency is good. A series of axial inductors, say 3.3,3,3, 4.7, 6.8, 10, 100uH will spread crystal and resonator coverage considerably.

There may be occasional frequency jumps but generally good stability, with for example the 14320kHz resonator covering the entire 20 metre band. Amazing things can happen at harmonic frequencies; with the bandpass filter at minimum capacity I have heard a local amateur loudly calling CQ on the 6 metre bands!

Please note that the tuning of the Bandpass Filter for maximum sensitivity is really sharp. I have been amazed at how sensitive the circuit has proved when correctly sharply tuned, in reasonable conditions on the HF bands especially the DX simply

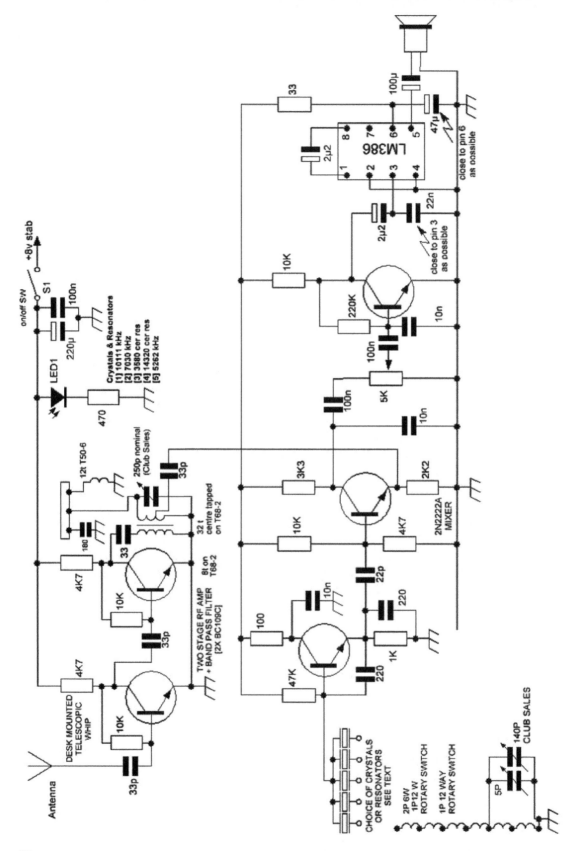

PART 3 RECEIVERS

pours in! The overall results are very encouraging for a very basic receiver without an RF Attenuator or AF Filter!

Construction should prove straightforward and results satisfactory if you take great care to keep all AF wiring away from the RF and Mixer stages of the circuit. Also, place the bypass capacitors at pins 3 and 6 of the LM3886 IC as close to the pins as possible. The LM386 enjoys singing unwanted tunes if such precautions are neglected! It is a great, cheap amplifier but can be hard to please if you fail to pay the fullest attention to avoiding unstable feedback

I hope the above gives you all the info you will require to build the CS5. If you hit problems, please feel free to contact me at keithcath@ranger144.fsnet.co.uk and we will discuss how to solve them.

Good hunting! You might even hear a 599 signal from OX3HR in Greenland, as I did the other day or CO3ARE in Cuba calling CQ at 579 strength on a completely dead 20 metre band – who knows!

QRP SCRAPBOOK

Building the "Sweeperino"
Nick Tile G8INE and Tony Fishpool G4WIF

This is not going to be a construction article but more an encouragement to go to the internet, download the considerable amount of material on this project, and have a go yourself. To publish the design in full would also take up most of Sprat. This article also describes a journey for the two authors as we explored new (to us) technologies and learned new techniques along the way.

The Sweeperino is a very useful Arduino based test instrument designed by Ashar Farhan VU2ESE, who also designed the well-known BitX transceiver.
It combines;
- A very stable, low noise signal generator from 4 MHz to 160 MHz without any spurs using the Si570 or the Si5351.
- A high precision power meter with 90 db range and 0.2db resolution using the AD8307.
- A sweeper facility that uses an Arduino to manage the process and communicate with the outside world.

The device generates a signal and then sniffs the output from the device that you have connected it to, typically, you can use it to be your antenna analyzer, plot your crystal or band pass filter through your PC.

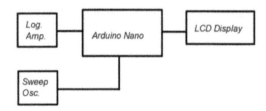

The block diagram (left) shows the 4 basic sections.
The device under test is connected between oscillator output and logarithmic amplifier input.

It isn't a complex piece of equipment, or very expensive to build (around 25-30 UK pounds), and it lends itself to modular construction, so you can get each of three major components working before combining them - making it manageable to build and much easier to troubleshoot.

Theoretically at least, the instrument will work from about 4MHz, the lower frequency limit of the Si570 (which is a programmable Oscillator), up to 100MHz, and if you are careful with the matching on the AD8307, it will be fairly flat. It could have a dynamic range of about 92dB (according to other articles) so careful construction is essential.

So what can you do with it?
You can test filters.
To the right is the sweep of a 37MHz low pass filter using VU2ESE's "Specan" software .

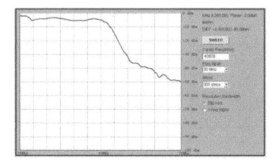

PART 3 RECEIVERS

The PC software is free from VU2ESE - although the Sweeperino also allows testing via the use of a physical frequency tuning control together with the LCD display.
Here is a photo of another filter being tested.

This is a 10MHz low pass filter, so the attenuation is shown as high – just as you would expect at 20 MHz.

During construction, we compared notes by email (as we live some 60 miles apart) and we wondered if the results that we were seeing were correct. So there was a slight diversion as we learned how to use "LTSpice". This free software will easily model filters and without a steep learning curve. There are some superb tutorial videos on the LTSpice website.

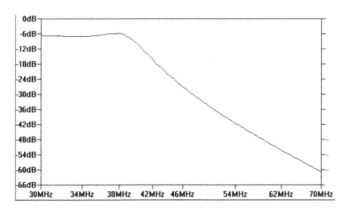

It is pretty similar to our sweeperino results, and we found that the use of LTSpice to be very useful for those occasions when you build filters from published articles - and you don't really know how they should perform. It takes just five minutes to draw the filter circuit and then you can see the predicted results as shown above.

Our construction techniques varied. While one of us likes to etch PC boards, the other builds "Manhattan Style" using "ME Pads" produced by Rex W1REX. These are available from club sales and of course direct from Rex in Maine USA.
Just when we thought that the project was finished we discovered that a Dutch group had designed an accessory providing enhanced facilities. Possibly the most useful was operation

QRP SCRAPBOOK

below 4MHz. This project was published in their June magazine called "Razzies" which is available free online. Downloading this article is highly recommended as it corrects a few errors in the documentation on VU2ESE's web page.

Just recently we've been experimenting with "return loss bridges". A very simple design was published by Jim Ford N6JF and was published in QST September 1997. You can still find the article online.

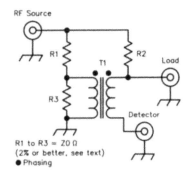

Using the Sweeperino with a RL Bridge to test a 20/15/10 metre trapped vertical gave results that were pretty close to an antenna analyser – I also tried a noise generator as a signal source – then used a TV dongle as a spectrum analyser to look at the signal coming back. The noise fall off in output (as frequency increases) has me now experimenting with monolithic amplifiers.

Hopefully this brief description will inspire a few members to have a go at this fascinating and very useful project. Significant thanks go to VU2ESE for documenting and sharing both this project and all his other designs.

References:
http://hfsignals.blogspot.co.uk/p/sweeperino.html
http://www.pi4raz.nl/razzies/ (you need the "juni" issue which is in two languages)
http://www.linear.com/solutions/ltspice
http://goo.gl/qRTGxk   (N6JF article)
http://www.fishpool.org.uk (G4WIF & G8INE construction notes)

Shown below, the G4WIF constructed return loss bridge.

Antenna Systems

Antennas Anecdotes and Awards
Colin Turner G3VTT 30 Marsh Crescent, High Halstow, Rochester, Kent ME3 8TJ
G3VTT@aol.com

Welcome to the Autumn issue of AAA. I must say a big thank you to all of you who have dropped me a line about their antenna experiments this Spring. Certainly one of those is Dan Reynolds KB9JLO who has had lots of success with his version of the evergreen W3EDP antenna. He's tried it with and without counterpoises and compared with his earlier trapped 80/40m dipole has found the EDP far superior. If you search on the web for the 'best length of a wire antenna' prominently 84 feet comes up as one of the more effective lengths. You can try the website 'www.hamuniverse.com' for the preferred lengths for antenna lengths up to 500 feet. In the meantime thanks to Dan for an enthusiastic report.

A 66' Top OCFD for 40 – 6 metres by Chris Baker G4LDS
Chris is one of my regular writers and he has been out again with wire, pliers and a soldering iron and he sent me extensive notes on an Off Centre Fed Dipole. Unfortunately due to space considerations I have had to edit some of the graphs but the basic information is below. Over to Chris.

66' top OCFD for 40 – 6 metres.

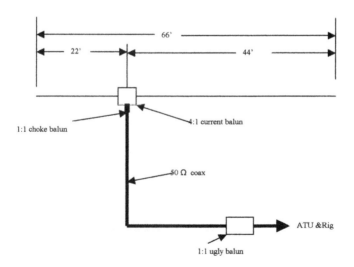

This aerial is a simple wire type based on the once popular multiband Windom which was first published in 1929 by Loren Windom W8GZ. This was a horizontal half wave on the lowest frequency of operation. This was fed via a single wire at a point approx 14% from the centre to the ATU. This point was chosen as a point of 600 ohms impedance. The

QRP SCRAPBOOK

theory being the centre of a dipole is approx 70 ohms and at the ends 2-3000 ohms so at a certain point it will be around 600 ohms impedance and this can be used to multiband the aerial by virtue of a fixed impedance point on harmonically related bands. The version I wanted was to operate from 40 – 10m and on the Internet there are several variants of this aerial called the 'off centre fed dipole' with a top of around 66 feet. Some are 66 feet others are 69 feet I notice. Due to its location, which will mean the feeder running down the house outer wall and the feed point just under the eaves of the house, I tried out the 66 feet top, not the Carolina Windom type, with legs of 44 and 22 feet.

Figure 1 (above) gives the basic design of the aerial and to get the match to allow multi band use the balun used was a 4:1 type. A 6:1 may be better but is a little more complex to build and heavier! Figure 2 shows the 4:1 balun The 4:1 current balun uses either two cores or one core with two windings on it. My first version used a single FT140-61 core with two separate windings each of around 200 ohms. I used enamelled copper wire of around 1mm diameter (0.8 – 1.2mm should be ok),

- Ensure the wire is straight and "kink free" by putting free end of wire in vice or around door handle and gently pull to remove any kinks but without stretching it!
- Cut off around 2 meters of wire and fold it in half to make a short length of 100-ohm transmission line.
- Using finger & thumb, carefully wind around 10-12 turns onto the core.
- Keep the wires flat don't let them twist or overlap
- Wind over about half the core
- I find that finding the half waypoint of the wire using a cable tie fix it to the core then wind half the turns one way anchor it using another cable tie then repeat on the other end of the winding. Thus making a 1:1 balun.

Repeat for the other winding. Then "buzz" using a multimeter through the windings to get the separate turns. Now comes the wiring up, on one side the two windings are parallel connected to give a impedance of 5Ohms, the other side are wired up in series to give an impedance of 200ohms. This is where most problems occur. To check it, use an aerial analyser, connect a 200 ohm resistor at the high Z side and, check it should be around 1:1 from 3 – 60 Mhz. if its more than say 1.5 check the balun windings wiring.

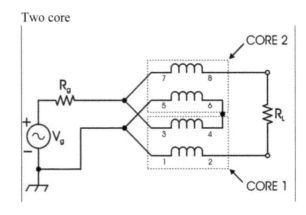

Figure 2 – the bare balun

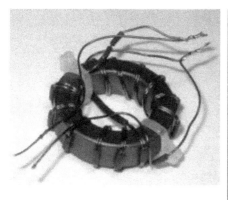

Figure 3 Balun in box.

Figure 3 gives some idea of how I weatherproofed the balun. I used the standard SO239 style connector for the coax feed and the output of the balun was connected to two bolts to enable the antenna wires to be connected to it. I used part of an ex chopping/cutting board trimmed to fit as the support for taking the strain of the aerial wire and anchoring. When connected to the wire I waterproofed the balun and SO239 coax connector with PVC then amalgamating tape. I then put modelling clay around two wire connectors! The set up is now a 1:1 balun at the aerial feed point consisting of around 10-12 turns of RG58 coax around 12" diameter then at the ATU and transmitter I made a 1:1 ugly current balun of a 40mm pipe with 30 turns of RG-58. I found that this aerial gave good matching on 14/28/50 MHz and an acceptable SWR on 40/10/18/21/24 with an ATU, *Chris included some excellent SWR charts which regrettably we do not have space to include.* A Mk II balun, *Fig 5* was also tried as a back balun using 2 type FT114-43 cores wound with 14 turns of 22Swg enamelled wire as in above notes. Each core was mounted above the other with a piece of paper between them using a couple of cable ties. I used different colour cable ties to show the start and finish of the winding. The wiring was as per figure 2 above. Both baluns gave a flat response of 1:1 from 2 – 60 Mhz.

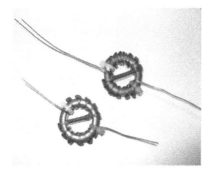

Figure 4. The twin FT114-43 cores wound with 14 turns

Two cores mounted and cable tied together ready for mounting into box.

Figure 5 – the two cores ready to wire into a balun box.

QRP SCRAPBOOK

AN EXPERIMENTAL LOOP ANTENNA for 7, 10 and 14 MHz
Duncan Telfer G8ATH/G0SIB(prev) e-mail: dtelf@talktalk.net

LOOP CONSTRUCTION

Whether expanding antenna options, having a change from long wires and their variants, or even starting from scratch, practical loop antennas are an interesting topic for investigation. So here is my account of one particular antenna that began with an excursion into plumbing...

A recent visit to a local metal scrap-yard yielded a few short lengths of 35mm diameter copper pipe. This tubing was trimmed to size and furnished with elbows from a local plumbing store to make a metre square loop. Because of what was available, the sides deviated slightly from 1m (97cm x 112cm). The pipe sections were de-burred and smoothed with wet-and-dry to expose bright copper. The elbows were sweated on by feeding cored solder into the joints while heating them from a blowlamp. A gap in the centre of one side was sleeved with 6 ins of plastic pipe, within which was fitted an inner sleeve 2.5 inches long to act as a spacer. PVC insulation tape was used to stabilise the structure and prevent sliding before mounting the loop tuning capacitor C1 on its wooden platform, which was secured to the outer sleeve with stout cable ties (**Photo 1**).

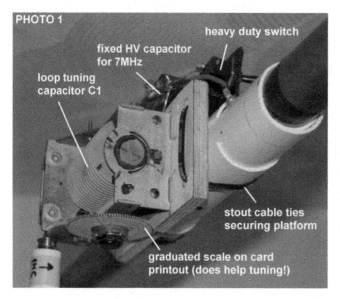

This tuning capacitor is a 150pF ex-transmitter high voltage single gang air-spaced type with reduction gearing, as tuning is very critical at loop resonance. Under these conditions, proximity to C1 is potentially dangerous as very high voltages are reached, even at QRP levels. Hence the length of plastic tube on the tuning knob. Across C1 is a switchable fixed 100pF capacitor, 3000V working, for when tuning the 7MHz band. The switch is also heavy duty and to avoid destructive arcing should not be operated while actually transmitting. If you don't need a fixed capacitor and have a suitable larger variable one, then fine. So long as its operational voltage is high enough – if unsure, DON'T test it with your transmitter or damage to the capacitor, and/or your rig, may result. **Ref. [1]** indicates expected voltages for loop antenna capacitors for given RF powers. For C1, a 'butterfly' split stator type or a high grade vacuum capacitor, may give improved results.

PART 4 ANTENNA SYSTEMS

Size 35mm diameter pipe was chosen because on-line loop calculators, e.g., Ref. [1] and EZNEC™ modelling predict a radiation resistance that is very low (~ 0.2 ohms at 7MHz). Even small resistive losses can degrade antenna efficiency.

Initially, the loop was mounted vertically, supported by a central fibreglass pole fitted in a home-made socket made from a short length of scaffold tube welded to a bracket bolted to an old swivel chair base (**Photo 2**).

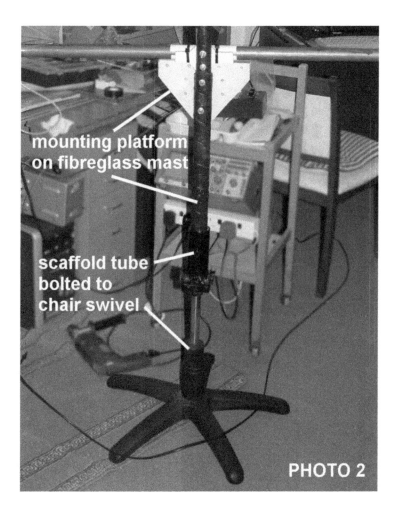

QRP SCRAPBOOK

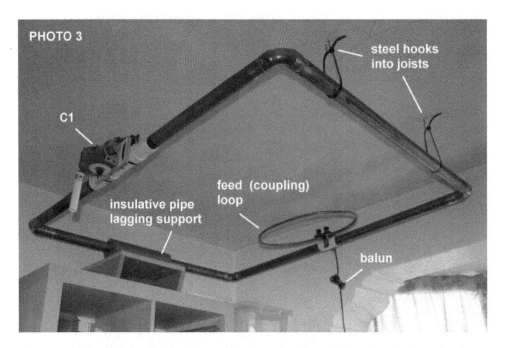

But now its favoured and safer elevated horizontal and omni-directional orientation is (**Photo 3**) near a corner of the ceiling, partly suspended by stout cable ties from hooks in the joists. [NB. Make sure the hooks ARE in the joists, not just the plaster! My joist detection device was supplemented by drilling pilot holes to check]. Preferably, tuning is performed first with an antenna analyser. It is helpful to have a scale of some sort on which to record or mark the position for minimal SWR. In use, loop tuning is performed for maximum noise (and/or signal) on reception, while re-checking SWR at the transmitter. But for safety, re-tuning should always be done with the transmitter off.

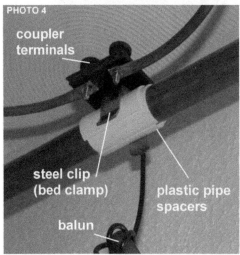

If your transmitter has a manual ATU, then switch it to 'direct' or 'bypass' path and simply use it as a reflectometer. It should not be necessary to invoke the ATU's filter components to get low SWR. That brings us to the next topic.

MATCHING UNIT (Photo 4).
The feeder loop clip-on assembly is a steel bed clamp, drilled to accept cable ties securing the co-ax cable and bolted to the plastic mount carrying the feeder loop terminals. Different feeder loops were tested until a reasonable match was found on all three bands.

PART 4 ANTENNA SYSTEMS

For experimenters, and in answer to those who might claim that the feeder loop size is 'non-critical', SWR figures appear in **Table 1** for different circular feeder loop sizes (10mm copper tubing). The measurements were made with the MFJ 259B and a short run (4m) of RG58U to each loop feeder bolted to the clip in position on the antenna.

TABLE 1.

Circumference (cm)	SWR at frequencies:		
	7.035 MHz	10.140 MHz	14.075 MHz
60	2.7	3.7	6.5
70	1.8	2.7	5.0
90	1.0	1.6	3.5
110	1.3	1.2	2.8
110 (oval)	1.5	1.0	2.3

The choice in this case was a distorted (oval) version of the 110cm feeder loop, which turns out to be 40cm x 25cm, with long axis parallel to the antenna tube.
By all means experiment further with your own feed loop geometries.

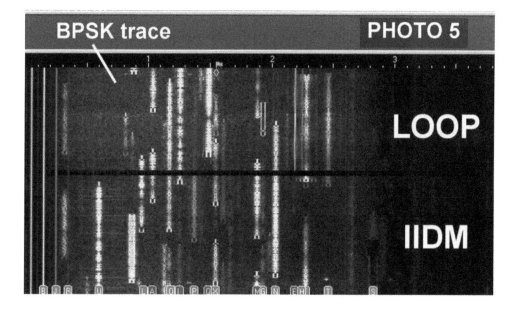

RESULTS

For NVIS (near vertical incidence skywave) propagation on 7MHz, the rotatable vertical loop was a respectable performer. But since I'd then be 'doubling up' with my multiband IIDM * (**Ref. [2]**), the preferred orientation for this loop antenna is horizontal, being 'out of the way' and omnidirectional at the same time. **Photo 5** shows

QRP SCRAPBOOK

the reception of BPSK signals at 10.140 MHz on cable swapping between the loop and IIDM antennas. Both antennas have effective heights of ~4.5m above ground. Many signal strengths appear similar, bearing in mind differences in far-field lobe geometries and feeder lengths. About 20m of co-axial connects the outdoor IIDM and less than 5m to the indoor loop antenna whose efficiency at this frequency is calculated in Ref. [1] to be 67%. Verdict: A worthwhile permanent feature of the shack. Good experimenting!

ACKNOWLEDGEMENTS

Thanks are due Brian, G0GSF for helpful discussions on loop antennas, Roy Lewallan W7EL for his excellent EZNEC™ software and to the efforts of other background workers in the literature, notably in **Refs. [3, 4 and 5]**. Google these titles for worthwhile further reading on small loop antennas.

* 'Inwardly Inclined Dual Monopole' - near equivalent to a low dipole antenna.

REFERENCES

[1] http://www.66pacific.com/calculators/small_tx_loop_calc.aspx
[2] 'Multibanding the IIDM' in Pat Hawker 'Technical Topics', RadCom, Feb. 2008.
[3] 'Small Transmitting Loop Antennas' by Steve Yates AA5TB
[4] 'An Overview of the Underestimated Magnetic Loop HF Antenna', Leigh Turner, VK5KLT 2009, 2010.
[5] 'Loop Antennas' by Glenn S. Smith, Georgia Institute of Technology.

LEGENDS for Photos.

Photo 1 – Main tuning capacitor, 150pF wide airspaced variable transmitting type, with switchable parallel 100 pF capacitor (mica, 3kV wkg) for 7MHz. Note the graduated scale (printout of a copied protractor image) and plastic insulator on the control knob.

Photo 2 – Portable floor mounting for rotatable vertical configuration, using an old chair swivel and section of scaffold pole to support a fibreglass mast.

Photo 3 – Elevated horizontal configuration for loop antenna, showing C1 opposite the feeder loop and its 50 ohm coaxial cable with current balun. Note the built up (plastic pipe covered) region to allow a firm fit for the clip-on feeder loop.

Photo 4 – Clip-on feeder loop assembly built on a suitably drilled bed-clamp bracket, incorporating a plastic terminal block for the feeder loop. Trailing coax is wound on a toroidal core to form a current balun.

Photo 5 – Comparing BPSK signals on the 30m band segment (s/w Digipan 2.0).

PART 4 ANTENNA SYSTEMS

Antennas Valves and Vintage
Colin Turner G3VTT
17 Century Road Rainham Gillingham Kent ME8 0BG
G3vtt@aol.com

Welcome to AVV for the Autumn. As promised there is something on antennas this time thanks to Ken G4IIB who lives in the Cumbria Lakes area and has sent these pictures and details of his home made magnetic loop antenna which is mounted upside down. The reason is the control box needs to be low down for maintenance purposes. The plastic container is water tight and Ken said he found some of the so called waterproof boxes from the electrical suppliers to be leaky, and expensive I guess.

A Simple Upside Down Magnetic Loop by Ken Marshall G4IIB

I live in a windy location in the English Lake District and I am a wheelchair user that values independence. So I need easy to build antenna solutions that are either robust or easy to take down when the wind picks up. In addition I am unable to climb ladders so something at or near ground level would be the ideal solution. After obtaining 3.5 meters of copper pipe 10mm in diameter for free I plumbed for a magnetic loop as I already had a reasonably spaced 10-130 pfd capacitor, the non-stop type, in my junk box. A magnetic loop operates on the magnetic rather that the electric component of an electromagnetic radio wave.

QRP SCRAPBOOK

After using this site to check dimensions etc:
http://www.66pacific.com/calculators/small_tx_loop_calc.aspx
I decided to make the loop 1metre in diameter, circular with no joints, giving a circumference of 3.14m. The loop would tune the 20, 17 & 15M bands giving a bandwidth of 33 kHz at 73pfd on 20 M, 61 kHz at 44pfd on 17M and 98 kHz at 33pfd on 15M.

Next I needed a motor to drive the capacitor. I searched eBay for a "12V reduction drive motor" and found one brand new from China for £7 including Post and Packing. This is the star of the show! It is 12V, reversible, draws about 40mA and has a high torque 3rpm gearbox and measures 4cm diameter and 3cm depth. I opened this up and can confirm that it has metal gears. The gearbox is integral to the motor. It will turn two fairly large capacitors i.e. with their spindles connected together with ease however I only needed the use one. This motor will still turn my capacitor with only 5V applied as at this voltage the motor completes one revolution in about 50 seconds. It is therefore useful for many remote tuning applications.

I made an 'L' bracket on which to mount the motor and connected it to the capacitor spindle via a plastic rod taken from the inside of a cheap Staples Biro. I housed the capacitor and motor, mounted on a piece of plastic chopping board, inside a plastic 18x25x9cm food container box. The size of the box you use would depend upon the physical size of your capacitor. The box has a water tight clip on lid and I use a cable gland to get the control cable into the box. The box is connected to both ends of the main loop via two stainless steel bolts (brass might be better but I had none). I drilled the holes in the box for the bolts slightly smaller than the bolt and tapped the holes with the bolts thus forming a water tight seal. Each bolt is secured with a lock nut and bolted to the ends of the main loop and connected via 2.5mm copper wire to each end of the capacitor. See image

Looking at the details of driven loop, for simplicity I opted for a shielded Faraday loop. This needs to be 1/5th of the circumference of the main loop so 20% of 3.14m = 62.8cm. Take a length of 50ohm coax (say 3m) and carefully remove a small section of insulation 63cm from the end of the coax cable and solder the centre conductor at the end of the coax to the braid 63cm in forming a loop. See image. Put a plug on the other end of the coax so you can connect to the shack via 50ohm coax. Make the connection watertight with self-amalgamating tape around the exposed braid. I use cable ties to position the driven loop at the top of the main loop. The mag loop is mounted on a 2.5m wooden pole and secured to my decking. The deck is 1m above ground so that the radiating part of the loop is 3.5m above ground. They work well in lofts too.

Because these motors are so good with loads of torque I have found that the control unit can be simple with no fancy electronics. I use a non-latching double pole double throw toggle switch to reverse the polarity on the motor shown in the diagram. In addition I use a variable power supply so that I can reduce the speed of the motor further by reducing the

PART 4 ANTENNA SYSTEMS

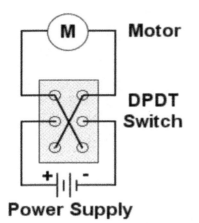

voltage (see the paragraph on motor details). I have placed ferrite clips on both ends of the control cable to protect it from RF. Another trick I have adopted is to tune to the frequency I wish to use on my SDR radio first and then tune in the loop. You can see the pass band moving along the SDRs spectrum display and it is quite easy to place the middle of the passband on your desired frequency then switch the antenna to your rig ready to transmit. The SWR is 1.3:1 - close enough for military work don't you think?

Most designs on the web, (but not all), show the capacitor at the top and the driven loop at the bottom I have opted to turn this around for accessibility purposes. I asked Colin G3VTT our club representative on antennas if this could affect performance and Colin said that having the loop "upside down" would have no significant adverse effect, he hinted that it just might mean that I am a little louder down in Australia. "Fair dinkum" as I have reached Australia with 200mW via WSPR.

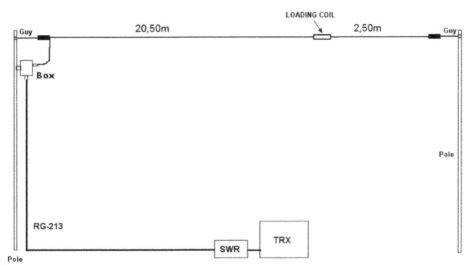

The 'Olandesina' 80 -10m Antenna by IK0IXI

Fabio IK0IXIhas been searching on the web and found this multiband aerial which looks like a useful project to me giving good results from 80 to 10m. Fabio explains, 'The antenna, the so-called "Olandesina" (little Dutch girl), is a project that I found on the web

QRP SCRAPBOOK

made by different Dutch OMs. In Holland this antenna is very common and there are some commercial versions too. This is not the familiar End Fed powered by a 9:1 transformer and tuner. It is a resonant antenna fed at one end by an impedance transformer having a ratio of 1:50. A half-wave dipole at resonance frequency shows the maximum current at its physical centre and its maximum voltage at the ends. Normally we feed our dipole "in current" by a coaxial cable of low impedance finding the right length to reach the minimum SWR. In this way, however, we get only a "dual band" antenna working as a half wave dipole on the fundamental and as a 3/2 lambda dipole (1.5 times the wavelength or three half-waves) on the third harmonic. In fact if you build a half-wave dipole for 7 MHz you got resonance on 21 MHz also. The Olandesina, thanks to the harmonic relationship between our bands, a half-wave wire will show high impedance at both ends on the fundamental and on all the harmonics. At these two end points there will always be a maximum voltage. In other words if the dipole is 20m long it shows high impedance at both ends on the range 7 MHz (half wave), 14 MHz (Wave Length), 21 MHz (the third harmonic - 3/2 wave - 1.5 lambda) and 28 MHz (two whole waves). In this case we built a resonant antenna on 4 bands that must only be fed in a certain way. To operate on 80m, we add a coil which acts as a load for this band. With only 23 meters total, a little bit more of a simple 40m dipole, we have an antenna that will allow operation on ranges 80-40-20-15- and 10m without a transmatch. The antenna has been built using a common copper wire for garden purposes.

Impedance transformer

In our "coaxial world" we are used to relating everything to the classic 50 Ohms. To feed this antenna by coaxial cable we must connect our low impedance cable to the high impedance of the Olandesina at its end. To do this we need a "step-up" impedance transformer which transforms the 50 ohm of coaxial cable to the high impedance of the antenna wire, say 2500 Ohm or more. We need a transformer with 1:50 ratio. Sometimes a 1:64 ratio transformer can be used. The UNUN transformer is closed in a plastic box being a waterproof type made by Teko. The input is a common coaxial socket SO-239 and the output is a porcelain insulator made by Johnson specifically for HF transmitters. The capacitor of 150 pF - 3 kV and is used to improve the impedance match on the 10m band. In fact on 28 MHz it seems to remove a certain inductive component on the transformer primary winding and the added capacitor improves the SWR. Removing the capacitor the SWR increases slightly, rising from 1 to 1.6 on 40 / 20m, while on 10m it rises to 3.0. The use of two ferrite core FT-240-43 allows to fed the antenna to the maximum legal power of 500W. I made many QSOs at the maximum power allowed by our legislation without problems.

Loading coil

The coil is made by winding 125 turns of enamelled copper wire, 1mm on a grey PVC pipe diameter of 32mm. The coil has an inductance of about 100uH.

RF Choke

To prevent any influence between the primary winding of the transformer and the cable shield is highly recommended that an RF choke is included just before the impedance transformer. It's made by using a toroid model FT-240-43 with 9 + 9 turns of

PART 4 ANTENNA SYSTEMS

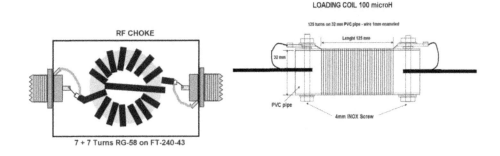

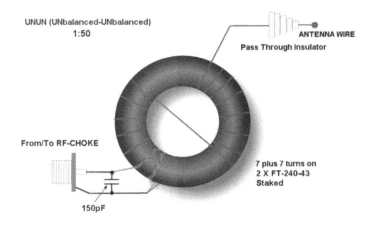

Details of 50:1 balun

coax like RG-58 and is enclosed in the same box containing the impedance transformer.

The antenna, with the measurements provided had no need for adjustment on the CW portions of the bands. However you can make small adjustments to SWR by adjusting the wire length. With a tuner, now built-in on all amateur radio transceivers, you can bring the SWR down to 1: 1.1 easily without compromising the antenna performance.

Performance
I work exclusively on CW with some output AM on 40m and 80m and the antenna performs well. I made many domestic and European QSO also DX is working well. On 80m it works but a 23 meters antenna is relatively small thing so the performance on this band is just usable. On 40m there is excellent performance with good reports and strong signals from all over the country again Europe and some DX QSOs. The SWR is 1: 1.3 on 20m, very good, tested for several months in the weekly sked I have with David VK3DBD. Both the regular HF bands and the WARC bands are usable with a tuner. Enjoy wire antennas as they are cheap and easy to build. Fabio IK0IXI

QRP SCRAPBOOK

Antennas Valves and Vintage
Colin Turner G3VTT, 182 Station Road, Rainham Gillingham, Kent ME8 7PR
g3vtt@aol.com

Welcome to the Summer AVV. It was very nice to work those of you who took part in the April Valve QRP Day and also those who regularly support the Monday evenings at 2000z gatherings on 80m. There is some more aerial information this time and I must thank the contributors. Please keep sending your ideas and experiences in to me at g3vtt@aol.com and I hope to work you on the bands. Chris Meadows G4KWH has been experimenting with magnetic loop antennas and wants to clear up a point or two plus give an overview of their uses.

MAGNETIC LOOPS G4KWH

To mag loop or not to mag loop - that is the question. There are many articles and web sites offering construction dimensions and building know-how for these aerials. So these few words will not elaborate on this aspect.

Having built some 12 or 15 such loops with a fellow amateur David G4KYX, the big test is to find qualitative measurements against other aerials, in other words 'how good are they'. This was not an easy job. First, to get a misconception out of the way: They do not receive only the magnetic part of the radio wave. A radio wave is an electromagnetic wave (the same as light from the sun) and exists in time phase and space quadrature and propagates through space in this way. A magnetic field can exist round a magnet and an electric field can exist near a conductor but they are not e-m waves, although electromagnetic in nature. I think what is meant by this first statement is that they are horizontally polarised in the same way as a horizontally mounted Yagi aerial. So mag loops are less responsive to urban vertically polarized wave interference. Sandy Heath transmitting mast is vertical but the aerial is based on the Slot aerial which is horizontally polarized hence all the aerials in the catchment area are horizontal Yagi's.

Mag loops are efficiently fed by a smaller loop about 25-30% of the main loop diameter. The feeding loop is often mounted at the bottom or at the top of the main loop. Loops may also be fed on the sides of the main loop and produce a largely vertically polarized e-m wave. Because of the very high voltages present at the capacitor, to transmit a power of more than a few watts a capacitor must have a distance between adjacent plates of 3 mm, whilst a valve superhet capacitor has a spacing of 0.75 mm. The latter can still be had at radio rallies for £2-3. Larger spacing capacitors are now priced in a range £40 to £90 and are not easily found.

Is all this effort worth the candle? If you live in a flat where space is a premium the answer is an unreserved yes. If you like building things then yes again. Compared to a long wire aerial mag loops pick up far less noise but the signal picked up is also less. Considering signal-to-noise ratios they are much the same as far as non-professional test equipment can measure. Mag loop signal loss is found to be between 6 to 10 dB when compared to a long wire aerial. But a mag loop can be easily rotated to null out certain (solar panels or power line communication?) noise. You cannot rotate a long wire. These aerials make a convenient portable device and are superior to a mag mount which has been found to add a further 3 to 6 dB loss. A 1 m loop fits in most cars and can easily work 40 and 20 m bands.

Chris **G4KWH**

PART 4 ANTENNA SYSTEMS

An adjustable loading coil for short loaded dipole antennas.
Ken Maxted GM4JMU, Newton Mearns, Glasgow, UK

In Sprat No 74, I published a 40m shortened dipole which seemed to attract some attention with amateurs having small garden plots. I wanted to put up this antenna in an Inverted-Vee configuration and make the best use of my 15m garden. The antenna I describe is a 14m version of the original but uses a coil design that permits easy adjustment of resonance.

Construction.

The antenna is constructed using SOTA-Beam green antenna wire (about 2.2mm diameter, PVC covered http://www.sotabeams.co.uk/antenna-wire-lightweight-100m/). Two 10.5 meter lengths were cut and terminated at a dipole centre insulator. I used a 1:1 balun but this is not essential.

The remote end of each wire was folded back through an end insulator and, using three small Nyloc cable ties was fastened back along its length for 300mm (this would allow for fine tuning if needed). At this stage you have a full-size 40m half wave dipole

QRP SCRAPBOOK

The shortening coils.

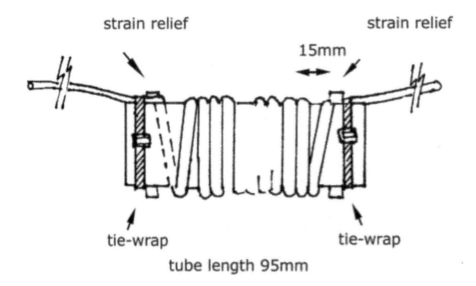

Coil construction

The coils formers are cut from 40mm diameter PVC waste pipe. However, the principal design feature is the fastening method: at each end of the 95mm length of pipe is drilled a pair of diametrically opposed holes of a diameter to suit the press-fit of a 50mm length of plastic or fibreglass rod. It is quite difficult to drill diametrically opposite so you may need a few attempts to get this right.

(I used 4mm fibreglass rod from B&Q for the strain relief rods but please note it is unpleasant to use as it sheds fine glass fibres when cut, and even rubbing the surface can cause skin irritation, and it is probably best cut using a Dremel type tool rather than a Junior hacksaw as the latter encourages fibres to split off). Thin wooden rod could be used, with the proviso that it is soaked well in paraffin wax. Nylon rod is probably the best choice.

The protruding rods serve as strain relief anchor points and all that is required to secure the coil is a large cable tie around both the former and the antenna wire,

PART 4 ANTENNA SYSTEMS

outboard of these strain reliefs. (See the picture) To adjust the coil turns or position all that is required is to slip the cable tie off the end, keeping it on the wire, remove or add turns and slip the same tie back on. The same can be repeated at the other end of the coil: to move the coil outwards and raise the resonant frequency, add turns to the outer end and take them off the inner end, make an identical change to the coil on the other leg of the dipole.

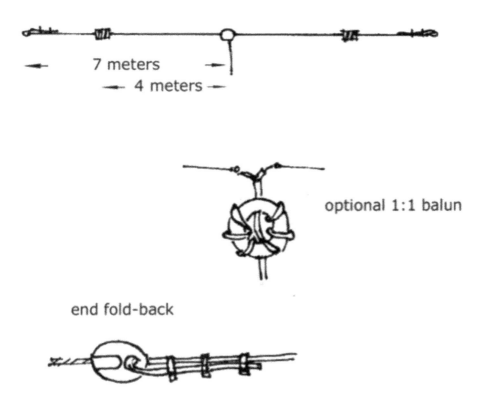

To make up the 14m (46') shortened dipole, measure out from the centre of the centre insulator a distance of 4 meters (put a small piece of tape on to remind you), hold the wire at this point to the middle of the length of the coil former and secure the wire to the former at the end of the tube with a tie-wrap. Make the first turn bending the wire behind the strain relief post and space the first turn spiral over 15mm of the length of the coil former.

From that point on, wind on 21 turns, close wound, and then make the final 23rd turn spiral out to pass in front of the second strain relief pillar towards the outer end of the antenna, keeping the centre set of coil turns close and under tension. Fasten at the end of the former with another tip-wrap.

The coil should be perfectly secure and when under tension the wire will pull on the strain relief pillars, keeping the coil tight but also protecting the wire from sharp angles of bend. This process is repeated for the other leg of the antenna. The distance from the centre insulator to the centre of the coil winding should be 4 metres.
With the tie wraps on the ends of the former, they can be slipped off in order to alter the turns on the coil and then slipped back on to secure the windings: do not fasten them too tightly in order to permit this.

Tuning up

The antenna should exhibit a resonance in or near the 40m band. If the antenna is too far LF the coils can be moved towards the ends by taking a turn off the dipole-centre side of the coil (lengthening the centre straight section) and adding one to the free end of the coil (making the straight section at the end of the antenna shorter by the length of one coil circumference). The ties are slipped off one at a time and slid back on to secure the new number of turns. I found that a frequency change of about 60kHz was made by adding a turn to one end and taking it off the other. If you wish to make a smaller change simply alter the length of the antenna end folds.

(To make the original half-size antenna, 40 turns can be put on the coils, the centre portion will be 2.57m and the outer section 2.27m. This will use all the space on the coil former- check a dummy winding before you cut a length of former and drill it. You will probably need a 160mm long former)

Performance.

Each half of the finished antenna will measure 4m to the centre of the coil and 7 meters overall length (with 300mm adjustment ends folded back). When supported with the centre at 7 meters above ground level and the ends 2.5m above ground the antenna had a resonant feed-point impedance of 48 ohms, so I get a very good match in the shack without an ATU. The useable antenna bandwidth is at least 100kHz, *the shorter version is closer to 30 kHz and its impedance is a little lower but still useable without an ATU.* The original half size antenna performed very well indeed, this longer one should be very useful as well, especially since it has a wider bandwidth, but radio conditions have been too poor recently to fully evaluate it. I have, howeve, regularly had good reports from the PSK reporter out to 1400 Km.

PART 4 ANTENNA SYSTEMS

Remote Antenna changeover relay
Bill G4KIH [bill.g4kih@gmail.com]

I have been using a home brew cobweb (see www.qrz.com/g4kih) for a number of years. This was located at the bottom of the garden on a 25 foot pole. The RG-213 feeder was fed back to the shack under the ground. An ideal situation you may say, until you want to add another antenna. A recently acquired IC-703 gave me a 4m capability, so the project of the remote changeover relay was born to allow a home brew 4m sleeved dipole to be mounted on the same pole as the cobweb.

The design idea can be found all over the internet, however a useful design for Sprat publication and can be used to switch two antennae down one coax.

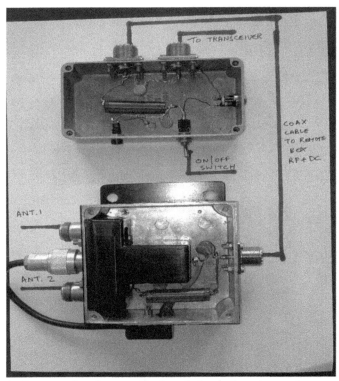

The project is in two parts, 1. Control box shack end, 2. Remote Relay box. The circuit is straight forward the only real decision is the choice of relay. For qrp good quality 12v car relays could be used, However I had acquired an aircraft antenna relay from some distant rally. So the choice of relay will influence the dc supplied and the led series resistor.

QRP SCRAPBOOK

Antennas Valves and Vintage
Colin Turner G3VTT
182 Station Road, Rainham Gillingham, Kent ME8 7PR
g3vtt@aol.com

There is some valve information this time plus an aerial and watch out for a new QRP activity period. Fabio IK0IXI, who is our representative in Italy, has provided some excellent photographs and drawings of another multiband aerial he has constructed. The photographs and diagrams give a comprehensive description of its construction.

Hi Colin, this is a simple 4 bands HF dipole that I built a couple of months ago and use daily. It works very fine and is loaded on 160m only and full size on other bands. Basically I have added two inductances on a 40m dipole. They act as an RFC for 7 MHz (and above) and as partial load for 1.8 MHz The 40m dipole works on 21 MHz in third overtone mode. For 20m a monoband dipole is connected at the same feed point. A 1:1 balun completes the dipole. Nothing else is required and it gives very good results on every band.

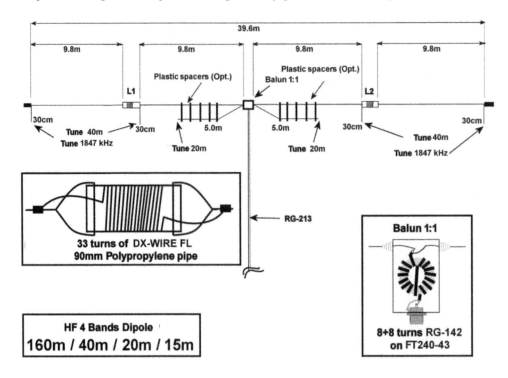

PART 4 ANTENNA SYSTEMS

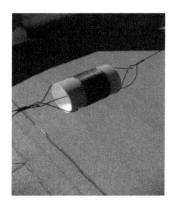

The balun construction next the loading coils for the dipole.

Is this a new Capacitor Free Trap construction method?

G3YVF has written and supplied a diagram about another method of trap construction he is considering using. So far as I know this has not actually been tried but it may well work. Is there somebody in GQRP who wishes to try this idea out? The idea stems from the tuning capacitor being removed and replaced by the capacity between the windings. I suspect that they may be some cancellation in the value of the inductance but this can be replaced by careful adjustment of the number of turns on the trap coil. The benefit would be cheaper construction, stronger construction, no expensive high voltage capacitor, more stable tuning if the turns of the trap coil are securely in place and lighter overall weight. Has he got something here? Please let me know how you get on if you try this idea.

Hole C is approx 1/3rd down the former. To trim for freq move wires a and b around the former taking fractions of a turn off at a time raises the resonant frequency of the trap. On a 1" 1/4 former and using approximately 6 red turns the capacity formed is approximately 45pF. All turns are tightly wound along side each other. As a starting point, and guide only, for a 14MHz trap, 1" 1/4 dia use 16 red turns and 4 or 5 blue turns. I use the same wire on the trap as the antenna so long as it is insulated wire!

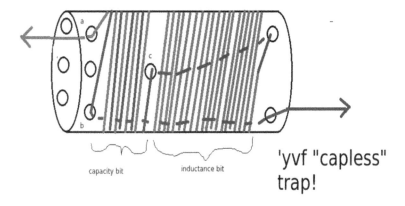

QRP SCRAPBOOK

Rig Safety
Dick Arnold, AF8X

One of the local hams, Paul, AA8OZ, ran into a problem while operating with his Elecraft K1. Paul uses a wire antenna held aloft by a kite and he recently had a problem caused by a static build-up on the antenna that zapped one of the components in his rig. After repairs were made Paul installed a 1K resistor across the antenna terminals in the radio to drain any static build up on the antenna. When Paul told me about this I thought my rigs should also be protected as I also operate portable using wire antennas with no protection.

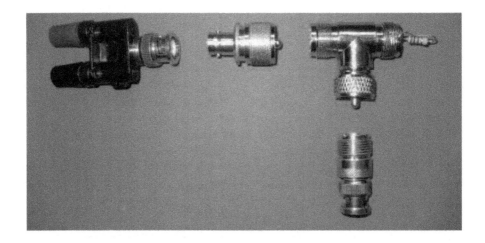

A little research told me that resistors between 1K and 10K would make a suitable drain resistor without affecting receive or transmit modes. I thought that I could install the resistor outboard of the radio and not have to find room in the innards.

I looked through my junk box and found a 2.2K resistor and a T-adaptor. I turned on the solder station and soldered the resistor from the center jack to the shell effectively bridging the antenna connections with the 2.2K resistor. For the K1 and KX-1 I needed an additional adaptor to adapt the SO239 to the BNC antenna connection on the rigs.

Now when connected to a random wire or available flagpole, I can be sure my rigs are protected from a static build up from either wind or nearby storm activity.

PART 4 ANTENNA SYSTEMS

A Doublet Experience
Ian Liston-Smith G4JQT

For a number of years I had a 42 metre random wire running north-south with the centre supporting pole mounted on the chimney reaching a height of 9.2 metres. At one end was the shack above the garage and the other end a tree. I had counterpoises running the length of the antenna and beyond, with three earth rods at the shack end.

Like most random wire antennas this arrangement could be matched on all bands from longwave to beyond 30 MHz, but I wanted to explore other antenna possibilities. Any replacement had to work from 160m to 10m, make use of the centre support and, importantly, allow coax feeder into the shack. This suggested a doublet-type antenna using open wire feeder from the centre to near the ground where it met the coax. But all the textbooks unhelpfully show the shack at or close to the doublet centre, with the open-wire feeder gracefully entering the building. My shack (like many others) is at one end of the antenna, and upstairs!

The obvious answer is a full-sized G5RV, where the coax can be routed behind sheds and between bushes, but even G5RV himself described his doublet-based antenna's limitations.

With many more bands now than it was ever designed to cover, I felt it was a serious compromise. Then I found the ZS6BKW antenna, developed about 30 years ago by Brian Austin who used computer analysis to design an optimised multiband doublet-style antenna.

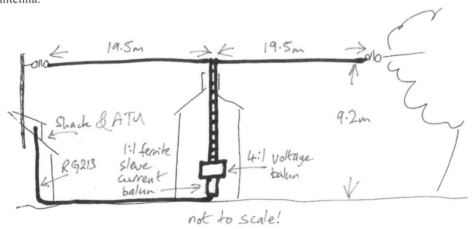

This looked ideal; it is a little shorter than the G5RV, but needs about two metres more height - depending upon the impedance of the ladder line. Unfortunately I doubt the present centre pole mounting arrangement could support an extra two metres, and the house is in a conservation area - and I don't want to push my luck!

QRP SCRAPBOOK

Back to square one.

It seems doublets are very versatile. They can be almost any length and height (although there are some dimensions best avoided), but do require a balanced feeder all the way into the shack. I've been told that routing this type of feeder is not as problematic as many textbooks suggest, but trailing it along hedges and behind sheds and between bushes for over 20 metres didn't seem like a recipe for success.

After a lot of web searching I found the comudipole (aka the coax-cable fed multiband dipole), attributed to PA2ABV. This is a doublet with a horizontal length of whatever is convenient, likewise any convenient ladder-line length to the ground, where it meets the coax via a balun. (I guess it's sort of G5RV/ZS6BKW, but made from whatever dimensions and space you have!) So I chose the comudipole, with home-made balanced line, via a balun attached the 24 metres of RG213 to the shack.

I found this antenna is described in the RSGB Handbook, 10th edition which also describes a suitable balun that can be wired 1:1 or 4:1. This exact article is available online in a pdf version of chapter 15 of the RSGB Handbook by just Googling "comudipole practical HF antennas".

During this research I discovered much about baluns of which I was unaware. The balun suggested in the RSGB handbook for this antenna is, I believe, a voltage balun. I later learnt that most informed opinion suggest this is not the best choice in a real-world antenna setup like this. But I don't intend to dwell on balun theory here.

I carried out some tests using a MFJ-259B antenna analyser at the shack end of the 25m coax run without any ATU. Whether using the 1:1, 4:1 voltage baluns or a ferrite sleeve choke balun (in effect a 1:1 current balun bought for a half-sized G5RV many years ago), there were no unmanageable variations in SWR on any band from 80m upwards. After assessing the numbers, the 4:1 balun gave the lowest SWR on most bands. The worst was 7:1 on 80m and best 1.5:1 on 17m with all other bands including 60m somewhere in between.

At about this time I became a WSPR user (using a Hans Summer U3 transmitter kit), and assumed this would be an excellent tool to test all of this. No doubt it would, given much time and statistical analysis. But unfortunately - particularly on the higher bands - there was so much minute to minute variations in measured S/N ratio (often varying by up to 20dB at any one station QTH in sequential two minute transmissions) any differences in balun configurations were not clearly apparent.

Even on 80m daytime groundwave I couldn't see a pattern emerging between the different balun arrangements.

PART 4 ANTENNA SYSTEMS

Then I found an interesting on-line balun article written by W7EL ("Baluns: What they do and how they do it") which gave me food for thought. My above tests indicated that the 4:1 voltage balun was the least worst balun solution for my setup (voltage baluns don't stop RF common-mode currents flowing on the coax outer). However the article said that a 4:1 current balun although being a bit awkward to make, especially to get consistent performance across all the bands, would be a better choice. However, the article suggests a 4:1 voltage balun followed by a 1:1 current balun was in effect a 4:1 current balun. That seemed like a good option, and the easiest way for me to do that was add my coaxial ferrite sleeve choke (my 1:1 current balun) to my 4:1 voltage balun's output just before it joins the RG213 coax cable.

Before doing this I operated my transmitter on 80m and went up and down the garden along the coax with an RF sniffer. Sure enough there were lengths where the coax was radiating. Adding the ferrite sleeve choke at the balun/coax junction noticeably reduced that radiation - I'd estimate by at least 6dB. A test on 20m didn't show as much difference. Adding this ferrite sleeve choke increased the SWR a little on some bands and reduced it a little on others, but maybe it will ensure that more of RF enters and leaves via the antenna now! I don't use enough power to ever get noticeable RF in the shack, so that wasn't an issue.

Of course there are a number of solutions to my requirements. For example I could use a fan dipole covering each band, but that's a bit conspicuous in a conservation area. Or a remote ATU at the bottom of the ladder where it meets the coax or mounted at the top of the pole to match the antenna and coax up there. Unfortunately remote ATUs are not cheap!

The present arrangement reduces the varying impedances of the doublet from band to band for easier matching by the ATU in the shack, which of course 'matches' the whole system, not just the doublet. Inevitably this means losses in the coax, but I hope my use of RG213 keeps that down to a minimum.

Conclusion

As far as I can see, my tests only prove that like most antenna systems, there are many variables to take into consideration and results just aren't clear cut. It's possible that my final configuration makes a more efficient antenna, but to be honest if I just connected the coax straight to the balanced feeder, I'm not sure if the theoretical deterioration in performance would make a noticeable difference!

Compared to the random wire it does seem quieter. Unfortunately, but unsurprisingly, it's worse than useless for 160m. Nevertheless loading coils or draping extra wire at the ends may resolve that. A future project perhaps…

QRP SCRAPBOOK

No Cost Traps
Richard Witney G4ICP

Traps seem to be making a come back and there are different ways to build them. This is my attempt at some 40m Traps made from scrap/recycled materials (i.e. zero cost), to be used for a 80/40m Trap dipole.

For me, the main criteria for Trap building are:
1) Sourcing suitable Capacitors
2) (Reasonably) Accurate and simple Trap tuning
3) Repeatability
4) Price (QRP style, cheap as possible, preferably nothing)

I heard that scrap offcuts of coax could be used as high voltage low value Capacitors. The capacitance occurs between the inner and outer conductors. A short length can be cut for the required C value, remembering to allow an inch extra or so for "cut and try" when tuning. I found the resulting Trap has high Q.

According to the data: (ref 1)
UR43 52 ohm cable has Capacitance of 29pF /Foot
so for 50pF cut 2 feet, which allows for some trimming.

(other coax cable types could also be tried)

For the 40m inductor, I used 18 turns of plastic coated single strand copper wire stripped out from some old "Twin and Earth" cable wound on 40mm diameter white plastic waste pipe

The coax "capacitor" is then soldered across the inductor exactly as you would connect a standard component, with the far end of the coax open circuit.

Tuning
I made a wooden jig to support the Trap clear of the bench. Resonance is checked with a GDO and referenced to an accurate receiver, in my case a SONY 7600D (remember those?) Other methods of trap tuning can of course be used.
Fine tuning is done by simply trimming the coax with cutters, starting low in frequency working up to 7100 kHz for a 40m Trap. Once tuned, the coax "tail" can be neatly folded, cable tied and stuffed inside the plastic tube.

PART 4 ANTENNA SYSTEMS

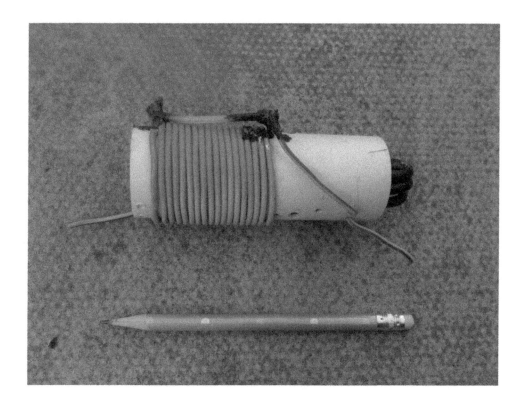

My red wire faded due to sunlight exposure over a 2 year service period. The black blobs are "liquid electrical tape" (ref 2) I had in stock and used for waterproofing the joints/ends etc. The coax is inside the tube and is just visible on the right.

I used the above method to make a pair of Traps for a 80/40m dipole following the standard aerial dimensions. It worked very well, in spite of being squeezed into a 30 foot square garden, but that's another story!
72 Richard G4ICP

Footnote:
It is worth mentioning that in order to achieve the textbook harmonically related multi-band operation of a 80/40m Trap dipole on 20/15/10m, certain criteria for L and C ratios must be met. My build was just for a 80/40m device so further experimentation is advised should this multibanding be desired.

References:
1: RSGB Radio Data Reference Book, Jessop G6JP 5th Edition p153
2: Liquid electrical tape is available from Richard G3CWI at Sotabeams.

QRP SCRAPBOOK

Trap Dipole Antenna for 5262 kHz and Beyond
John F Alder G4GMZ   johnalder1@btinternet.com

Having recently built an 8W TX for 5262 kHz, thoughts turned to an antenna built to match it. The principal target was a wire antenna resonant at 5262 kHz and any other frequencies that could be achieved would be a bonus. The junk box contained two well constructed antenna coils from a now deceased "Half-G5RV" antenna and these were measured to show near identical inductances and dimensions shown in Table 1.

Table 1: Dimensions of Trap Loading Coils	
Former	38mm dia. x 100 mm long x 1.5 mm thick black plastic tube
Coil	60 turns 1 mm dia. enamelled wire over 75mm; varnished over
Inductance	82 µH measured with a Peak Electronics LCR bridge
Connections	Screw clamp onto wire

Extensive reading and particularly the web page of KC9RG: http://www.i1wqrlinkradio.com/antype/ch36/chiave20.htm who refers to *"ARRL Antenna Book (1988, pp 6-6, 6-7, the "Loaded Antennas" section of Chapter 6, "HF Antennas for Limited Space")"* set me up with an idea of the initial dimensions with which to work. The two inner lengths of wire from the Half-G5RV were still available and these were used in the initial set up as in Figure 1.

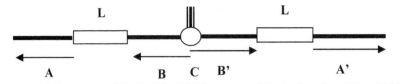

Figure 1: Dimensions A&A' are from end of formers L to insulators; B&B' from centre connector C to formers. The actual length of wire cut was slightly longer due to connections [*see text*]. The coax was 10m RG58/U terminated in connector PL259. Antenna wire in final construction was PVC coated 32 x 0.2 mm [1.0 mm^2] tinned copper [SOTABEAMS UK]

An MFJ 269 Antenna Analyser was used, which measures and displays the resistance [*R /ohm*] and the reactance [*X /ohm*] of the impedance, and SWR of the antenna & feeder assembly at the set frequency. Readings were taken after varying the outer elements A&A' between 3.0 and 0.95 m with the inner elements B&B' at 4670mm. The antenna was strung up at 1.5m above ground level [*agl*] and wood-propped centrally to be nearly horizontal

The first set of readings recorded the frequency up to 15 MHz at which the SWR fell between 1.0 and 1.1; the results are plotted in Figure 2 for the range 3 to 5.5 MHz. It was noted that a resonance [R~40; X=0; SWR= 1.1] occurred at 15 MHz for all values of A&A'. It was suspected, rightly as it transpires that this was a property of the inner part of the antenna LBCB'L. Such behaviour may be obvious to some but it was a revelation to me how well the inductors isolated the inner from the outer part of the antenna. It pointed the way to optimise the usefulness of the construction.

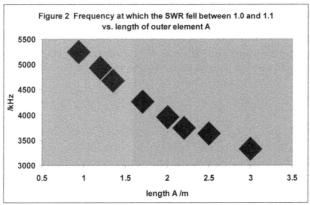

The next step obtained readings of the R, X & SWR at the QRP calling frequencies: 3560, 5262, 7030, 10116 & 14060 kHz; frequencies above 15 MHz are not reported here. The results are in Table 2.

Table 2	Values of R/X/SWR read from the Antenna Analyser for decreasing lengths A&A' at the chosen QRP calling frequencies; B&B'=4.67 m				
A&A'/m	3560 kHz	5262 kHz	7030 kHz	10116 kHz	14060 kHz
3.0	30/17/4.3	506/0/3.9	14/0/4.0	638/0/3.1	170/0/2.2
2.5	66/31/2.7	514/0/3.9	14/0/4.0	636/0/3.1	181/0/2.3
2.2	33/12/2.3	507/0/3.7	15/0/4.1	704/0/3.1	169/0/2.3
2.0	46/15/2.7	537/0/3.9	15/0/4.0	634/0/3.0	208/0/2.4
1.7	48/30/3.5	591/0/3.9	15/0/40	655/0/3.0	199/0/2.4
1.5	38/44/3.9	671/0/3.9	15/0/4.1	705/0/3.1	203/0/2.4
1.35	36/44/4.1	769/0/3.9	15/0/4.0	722/0/3.1	205/0/2.5
1.2	34/43/4.2	915/0/3.6	15/0/4.1	713/0/3.1	209/0/2.5
0.94	36/43/4.2	78/0/1.2	No data	No data	No data

A rather sharp dip occurs at 5262 kHz when A&A'=940 mm, whereas the parameters at other frequencies stay relatively unchanged by altering lengths A&A'.

The next step was to alter the 15 MHz resonance to 14 MHz for a better match before optimising the length of A&A' for good resonance at 5262 kHz. Using a few rough calculations scaling the resonant length at 15 MHz to a prediction for 14 MHz while maintaining A&A' at 940 mm gave the results in Table 3.

QRP SCRAPBOOK

Table 3 Values of R/X/SWR read from the Antenna Analyser for decreasing lengths B&B' at the chosen QRP calling frequencies; A&A'=940 mm

B&B'/m	3560 kHz	5262 kHz	7030 kHz	10116 kHz	14060 kHz
5.13	21/45/3.7	255/0/2.3	13/0/4.0	520/0/2.7	23/0/1.9
5.00	26/39/3.9	148/0/1.9	13/0/4.2	453/0/2.5	35/0/1.2
4.88	10/37/3.9	48/4/1.0	6/10/4.5	276/0/2.6	51/6/1.1

The results showed favourable SWR at 5262 and 14060 kHz together when B&B'=4.88 m. It was remarked also how the values at 10116 kHz were within the range of matching through an ATU. My practical experience is that values of R<<50 and of X>>1 are not conducive to good matching through my ATU whereas quite high values of R are OK as long as X~0.

The final iteration was now undertaken holding B&B' at 4880 mm and varying A&A' between 940 and 880mm showing a good match when A&A'=940 mm. The results showed principally the scatter of results around the resonance at 5262 kHz, varying between R=58~142 X=0 and SWR= 1.1~3.0. It was decided that further optimisation would be pointless until the final construction was made.

The structure was rebuilt using all new antenna wire [see legend to Figure 1] and strain relief built in using cable ties. The antenna was put in place with the ends at ~4.0 m agl and the centre at 3.5 m agl using porcelain end insulators and 2mm woven nylon guying. The final measured dimensions were A&A'=920 (1000*) B&B'=4850 (4880*); L&L' 100, all /mm. The values* in brackets are the actual length of wire and the smaller value is the final measured length, the differences being due to the strain relief loops. Clearly my allowances for loops were inadequate! The optimisation process should really be repeated to overcome this discrepancy in order to have a perfect impedance match with no ATU; perhaps one day!

The tests were repeated near to the QRP frequencies and results shown in Table 4.

Table 4
R, X & SWR of Final Antenna near the QRP Frequencies

/kHz	R	X	SWR
5181	51	23	1.5
5190	47	12	1.2
5200	45	0	1.1
5210	50	3	1.1
5220	66	11	1.4
5230	84	3	1.7
5240	120	2	
5250	151	0	2.3
5262	230	0	2.5
5271	300	0	2.7
5284	420	0	2.8

5 Test Equipment

A Quick and Easy GPS-Locked Frequency Standard
Dr Andrew Smith G4OEP (aj-smith@blueyonder.co.uk)

The more you become involved in home construction and design, the more necessary it is to have stable, reliable and precise frequency measuring equipment, both to evaluate home construction products, and to ensure compliance with band plans and licensing requirements. Some amateur modes, such as qrss, require extreme stability and precision and many qrss operators use rubidium standards to control their rigs, although an ovened crystal will give adequate stability while leaving calibration uncertain. So a quick and easy frequency standard which allows reliable and precise frequency calibration is highly desirable. In the UK the LF station *MSF* at 60kHz is intended to be a national resource for frequency and timing calibration. I have a system which phase-locks a 2MHz VXO to the 60kHz carrier, the 2MHz output serving as a timebase for a home brew frequency counter with 1Hz resolution in the HF region (see G4OEP website http://g4oep.atspace.com/).

Unfortunately MSF is too often off-air, so GPS therefore seems a useful alternative. Most easily-available GPS receiver modules include a 1Hz output with a claimed precision of parts in 10^{11}. Technically, locking an HF oscillator to a 1Hz standard seems a tricky task likely to raise problems of lock-time, phase noise, etc. But the technique illustrated here has proved to be surprisingly quick and easy, and although I have not evaluated performance rigorously, it is clear that this exceedingly simple technique is useful. Lock can be achieved in about 3 or 4 minutes from a cold start and there is no perceptible phase wobble relative to a free-running crystal even when compared on a dual-channel scope with a 20ns/div timebase. The GPS-locked system is more than adequate as a timebase for a frequency counter, and the purity of tone of the harmonics in the HF band indicates that it is also good for even the most demanding of amateur requirements in other applications. In short, the technique is quick, simple, and highly recommended.

A PIC 16F627 is used to lock a 2MHz crystal although other frequencies can also be used. The internal T0 counter and the pulse-width modulation (PWM) module of the PIC are exploited. The technique is as follows:

The PIC's internal clock is derived from an external 2MHz VXO, which is tuned over a narrow range by a varactor diode. Inside the PIC, the clock input is divided by 4 to provide the CPU clock. T0 is an 8-bit counter which, when enabled, increments at the CPU clock frequency and creates an interrupt each time the counter overflows. The interrupt service routine preloads the counter with a fixed number, (nominally 6) every time T0 overflows so that successive overflows occur every 250 CPU clock cycles, or every 1000 VXO cycles. The result of this is that interrupts occur every 500us (2kHz). Because the T0 counter runs continuously, its contents represent a running total of half-milliseconds, mod 250, and this is a continuous count of the VXO cycles, divided by 1000.

QRP SCRAPBOOK

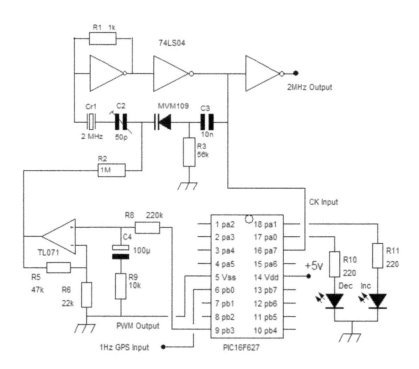

The GPS 1Hz pulse is applied to the external interrupt input to the PIC. These interrupts are not enabled, but the interrupt flag is monitored in a loop in the main program, and each time a rising edge of the GPS 1Hz occurs, the contents of T0 are read. Because T0 is cumulative, its contents can be interpreted as a phase error. Since T0 counts mod 250, and 500kHz is an integer multiple of this, T0 will have the same value each second if the 2MHz VXO is exactly on-frequency. If the VXO is running too fast, the read value of T0 will steadily increase, while if it is running slow, the T0 count will steadily decrease.

To complete the frequency control loop, the value of T0, as read each second, is written to the register of the PIC's PWM module which controls the period of the PWM waveform, the high phase being fixed, set in the initialisation routine of the software. The PWM output is filtered to recover its dc component, amplified, and then applied to the VXO as a frequency control input. If T0 increases, for example (due to the VXO running fast) the average PWM output voltage falls, the capacitance of the varactor increases, and the VXO frequency is corrected in the required downward direction.

The system could lock to any frequency (f) which satisfies the equation $f = 1000*Ni$ where Ni is an integer - the number of interrupts each second (2000 in this application). The nearest alternative frequencies are thus 2MHz +/- 1kHz. A crystal will not drift by as much as 1kHz, so there is no possibility of a false lock. Clearly other frequencies can be generated by simply changing the crystal, provided the target frequency is an integral

PART 5 TEST EQUIPMENT

number of kHz. The system has been tested with a 4MHz VXO, and I can confirm that it works well at this frequency without any change in the software. The capture transient can be expected to be slightly different at different output frequencies.

The loop filter consists of a 100uF tantalum capacitor with a 220kOhm resistor in a first-order LPF configuration.

A 10k resistor is added in series with the cap to increase damping. Damping is not optimal, but is adequate (see graph; this shows the locking transient following a completely cold start, including the capture time of the GPS receiver).

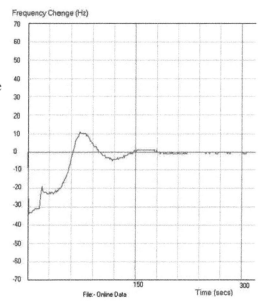

An amplifier with a voltage gain of about 3 is included in the loop. The op-amp is powered from a MAX232 chip which provides +/-9v supplies from a single 5v input. This allows a useful increase in the range of the control voltage (from about 5v maximum from the PWM module, to about 8v).

The software can be downloaded using the link indicated below. Astonishingly, the basic control system uses just 49 bytes of assembled opcodes. To my mind this is amazingly simple and efficient. The component count could be reduced by omitting the opamp, and by using the PIC's internal oscillator amplifier instead of the TTL inverters, but I leave these ideas for experimenters to explore.

Additional information and updates can be found here:
http://g4oep.esy.es/gpspll/gpspll.html

Software can be downloaded here:

http://g4oep.esy.es/gpspll/gpspll.zip

I would like to thank Hans Summers G0UPL for generously providing the SKM52 GPS receiver used in this project.

A WSPR Frequency Calibration Tool
John S. Roberts, G8FDJ

Accurate frequency setting is an essential requirement for anybody wanting to use WSPR. The HF WSPR bands are only 200Hz wide, so an external method of accurately setting the receive frequencies is a useful aid . The method described below takes advantage of the popular AD9850 DDS module driven by an Alduino Nano microcontroller. A total of ten WSPR centre frequencies ranging from 160m to 10m are sequentially activated and the signal for a given band displayed using a computer sound card and the free spectrum analyser software "Spectran V2" [1]. The shortest repeat time for a signal is just a few seconds, so the waterfall display provides a series of pulses that can be easily distinguished from WSPR data. The screen shot of figure 1 shows how Spectran presents the output from this unit. The aim is to align the signal with the 1500Hz cursor position, which is the centre of the 200Hz wide WSPR band. The receiver bandwidth should be set for USB, so the method still requires the tuning accuracy of a few Khz. (The 1Khz CW signals shown on the waterfall display are an artefact of the USB sound card.)

An Arduino sketch has been uploaded to the G-QRP web site and is essentially an extension of the AD9850 DDS driver programme written by Andrew Smallbone [2]. A 16x2 LCD display flashes the WSPR transmit frequency as the signal is transmitted. There are two scan rates set by a toggling pin D7 of the Nano; either 2.5 seconds/frequency (D7 = HIGH) or 0.3 seconds/frequency (D7 = LOW). The longer scan allows easier reading of the display and Rx tuning while the shorter period aids the Spectran waterfall display.

The DDS module will have improved stability if the 125 MHz oscillator module is thermally insulated. The photo of the completed test source shows a black silicon rubber thermal insulator made from "Sugru"[3]. Calibration was undertaken using the 10m WSPR centre frequency of 28.126100 MHz by comparing the spectran display with an accurate cw source. Corrections were made by trimming the 125000677 frequency shown in the Sketch at line :-

int32_t freq = frequency * 4294967295/125000677;

A calibration accuracy of about 10Hz was possible using the Spectran cursor and 10M frequency.

There are usually enough signals from the module to be detected without an aerial. However, the 80M and 160M bands can benefit from a short piece of wire on the SINB output of the DDS module.

Pin connections between the Arduino Nano and both the lcd display and DDS module are attached to the Arduino Sketch. A separate regulated supply was used to power the DDS module and lcd display, rather than rely on the 5v output from the Nano. Construction was based on "DAR-tec" matrix board and 0.1 inch PC connectors to support the modules.

[1] www.weaksignals.com
[2] www.rocketnumbernine.com
[3] www.sugru.com

PART 5 TEST EQUIPMENT

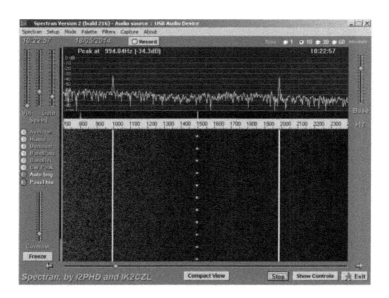

Figure 1

Completed unit with thermal insulation on 125 MHz DDS module

QRP SCRAPBOOK

W1FB MEMORIAL ENTRY
Signal Injector and Tracer
Peter Howard, G4UMB, 63 West Bradford Rd Clitheroe Lancs BB7 3JD

Here is a project which can be made in a weekend and is useful for testing projects or general fault finding without resorting to expensive instruments. The tracer is basically a two transistor amplifier with a diode detector. The injector is a multivibrator oscillator that generates a signal that is rich in harmonics. I saw a circuit similar to this many years ago in a Practical Wireless Take 20 feature; but that circuit would deafen you on inject if you didn't remove the earpiece!. The case is a piece of 20mm diameter wardrobe clothes rail. It can simply be made only as an injector by deleting the earpiece, switch and diode.

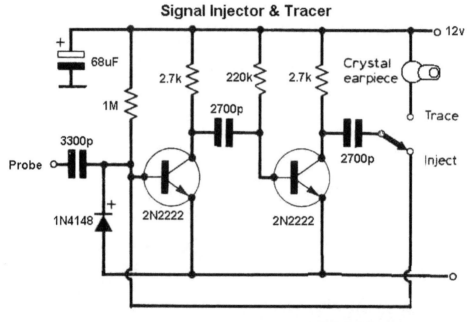

PART 5 TEST EQUIPMENT

VFO and Signal Generators using the Si5351A chip
Hans Summers, G0UPL, http://qrp-labs.com, hans.summers@gmail.com

The SiLabs Si570 chip has been popular with homebrewers for a while. It comes in several variants and can synthesise an output frequency from 10MHz to 1.4GHz. Two issues with this IC are 1) frequency coverage limited to above 10MHz, and 2) it is rather expensive!

The relatively new Si5351 chip is a cousin of the famous Si570. It requires a crystal or oscillator as reference for the internal PLL, unlike the Si570. There are several types of Si5351, with VCXO input, External auxiliary clock input, and with 3 or 8 independent outputs. We'll just consider the smallest/simplest/cheapest type, the 10-pin MSOP package Si5351A. This chip costs not much more than $1 and has gained increasing popularity among amateur radio homebrewers in the last year. The low cost, good frequency stability and spectral purity, multiple outputs and wide frequency coverage make it a great alternative to other synthesisers such as Direct Digital Synthesis (DDS).

The Si5351A can produce three separate outputs, each of which can be configured to any frequency from 3.5kHz to 200MHz with very high precision (a fraction of a Hz). In fact, my tests show that even though the datasheet says 200MHz is the maximum, it actually will go up to around 300MHz! The reference crystal should be a fundamental mode crystal in the range 25-27MHz, connected straight to the chip's internal oscillator.

This diagram shows the internal blocks of the Si5351A chip. With only 10 pins, there's not much to go wrong! The chip requires a 3.3V supply at VDD and VDDO, with decoupling capacitors close to the chip. The crystal (25-27MHz) is connected across XA and XB pins. Although the datasheet recommends certain pricey crystals, in fact any old common inexpensive crystal should work fine.

The chip's numerous internal registers configure its operation and output frequencies. They are programmed via the common I2C serial protocol (pins SDA, SCL), which is also known as Two Wire Interface (TWI) in Atmel microcontroller datasheets. Programming the registers using I2C is a job for a microcontroller. Many microcontrollers have built-in I2C peripheral; if not, it is not complex to implement a "bit-banged" I2C in software.

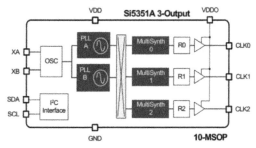

Conceptually, the chip consists of a reference oscillator (25-27MHz crystal), which is multiplied up by a Phase Locked Loop (PLL) to an internal Voltage Controlled Oscillator (VCO) frequency which is in the range 600-900MHz. That frequency is divided down by a "MutliSynth" divider. Each of the three outputs has its own Multisynth divider, and there are two PLLs. The outputs also have a further divider stage (labelled R0, R1, R2 in the diagram) that divides by a power-of-2 between 1 and 128. The PLL multiplication and the MultiSynth division are both fractional, which allows very high accuracy of the output frequencies. However the

QRP SCRAPBOOK

datasheet recommends for lowest jitter (phase noise), the MultiSynth division ratios are restricted to even integer values. In the middle is a cross-switch that allows flexible routing of the PLLs to the MutliSynth dividers. The outputs are 3.3V pk-pk squarewaves with 50-ohm impedance.

The Si5351A chip is surface mount and really tiny, just 3 x 3mm. The 10 pins have a spacing of only 0.5mm! I have successfully "ugly" mounted these chips with tiny wires but it's not an easy job,. Avoid coffee or alcohol for at least the prior 24 hours! Fortunately, a number of vendors now produce "break-out" boards or kits, with the Si5351A chip already soldered to the PCB. This makes using the chip in your projects very easy! See for example QRP Labs http://qrp-labs.com/synth Si5351A Synth kit, $7.75; Adafruit Si5351A breakout board, $7.95 https://www.adafruit.com/products/2045

This photo shows the QRP Labs Si5351A Synth kit. It has onboard 3.3V regulator, and two BS170 transistors acting as I2C level converters, so it is suitable for interfacing with 5V microcontroller systems such as the popular Arduino. The circuit diagram is below.

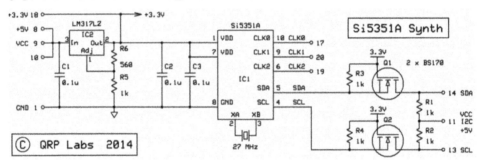

QRP Labs also produce an Arduino Shield for the Si5351A Synth kit, with an onboard QRPp PA and sockets for a plug-in LPF. Example sketches demonstrate how to configure the Si5351A registers for a desired output frequency. These examples will be useful for anyone wishing to write their own code to control the Si5351A, see http://qrp-labs.com/synth/si5351ademo.html

A nice VFO/Signal Generator using the Si5351A Synthesiser can be made by the addition of a microcontroller, display, and rotary encoder. Rotary encoders are available in mechanical types with actual mechanical switch contacts, or much more expensive optical encoded types with more "clicks" per revolution of the knob. In both cases however, there are two outputs, which are offset by 90-degrees phase. For each "click" of the shaft rotation, the two outputs pass through all four phases. By detecting the order of the phases seen on the two output signals, a microcontroller can determine the direction and speed of rotation.

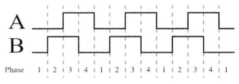

(Continued on page 142)

PART 5 TEST EQUIPMENT

5 in 1 Tester
Tests NPN - PNP - FET - XTL - LED
Peter Howard G4UMB 63 West Bradford Rd Waddington Lancs

5 IN 1 TESTER

This is a basic 5 in one testing board to give an indication of whether a transistor is working. It is also useful to check that a crystal works on frequency in conjunction with a radio receiver and to check LED's and their polarity.

To begin ensure that all the transistors are good ones and the LED lights. Then exchange whichever the one is you are going to test. Finally remove the crystal to ensure that the LED goes out to prove that the transistors turn off and are not just shorted . The circuit works by the FET Pierce crystal oscillator oscillating and then this signal is rectified which is able to switch on the next transistors to light the LED. .

felt no need to put this in a box so it's a cheap test rig and only takes an hour or two to make. Purposely made with three different types of transistors it will work with JFET positive types. Such as 2N3819, BF245 , J113 etc. and small NPN transistors like 2N2222, BC108 etc and PNP types BC212, BC558 etc. The wire ended crystals which the club sell fit nicely into the IC socket It's better to use the turned pin type IC socket with the round holes which I have cut into sections. It's helpful to label the sockets which as you can see from the picture they are done in the sequence EBCE and DSGD to accommodate all transistor body connections.

This is an improved version of an older design because each stage is DC isolated . The unit is built on stripboard.

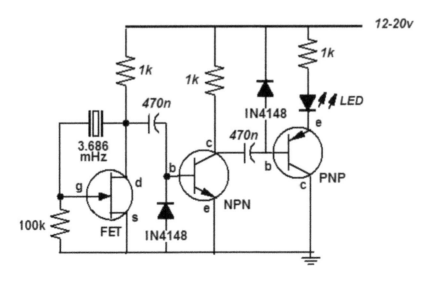

(A photograph of the 5 in 1 Tester is on the following page.)

QRP SCRAPBOOK

The 5 in 1 Tester

(Continued from page 140)

Almost all examples of published rotary encoder firmware I have seen, rely on either resistor/capacitor circuits for switch debounce, or timers in the microcontrollers. I tried numerous examples and was satisfied with none of them. They miss rotation clicks or still suffer false outputs due to switch bounce. All too often, the quality of the rotary encoder itself is wrongly blamed, particularly if it came from eBay or China! Cheap ones are fine.

The solution is elegant and simple. I wonder why it isn't done this way more often! The microcontroller need only look for a phase transition (see above diagram), and when it occurs, store it. The processor should then ignore any transition back and forth between those two phases, and only react to the adjacent phase transition, which will tell it which way the shaft has been rotated. For example suppose the processor sees the change between phase 1 and phase 2. It should then ignore any more changes 1->2 and 2->1 which could be due to switch bounce. When it sees 2->3 or 1->4, it will know whether the knob has been turned clockwise or anticlockwise. All with no switch bounce problems!

QRP Labs offer a VFO/Signal Generator kit including Si5351A Synth, 16 x 2 blue-backlit display, rotary encoder and programmed microcontroller. The kit produces a fixed (configurable) output that could be used as a BFO for example; and a rotary-encoder tuned variable output. Both cover the full available range of the Si5351A, i.e. 3.5kHz to over 200MHz. Connection of a GPS receiver with 1 pulse per second output permits calibration and continuous correction of the frequency. A very accurate VFO or Signal Generator for the shack! Please refer to http://qrp-labs.com/vfo for circuit diagrams and information.

RF Voltage Source Test Generator
Mark Dunning, VK6WV   marcommd@tpgi.com.au

RF Voltage Source Test Generator

I wanted to test some ideas and diodes for RF probes. I soon got tired of setting up my signal generator and I was not 100% confident that it was a sine wave anyway. I looked in my junk box and whipped up a test generator. It is based upon a surplus 4.8MHz CMOS oscillator I acquired some time ago. This sort of module seems widely available now quite cheaply. Just be careful to check what the supply should be. They used to all be 5V but 3.3V or lower is becoming more common.

The square wave output was filtered back to a sine wave and a stepped attenuator and termination load for the filter provided using screws as test and mounting terminals. The choice of 4.8MHz was based upon what I had on hand and the fact that QRM from it would not interfere with any background listening while I was tinkering. Another frequency would be fine but the filter would need to be scaled to suit.

I have been surprised at the different performance of diodes of nominally the same part number from different suppliers. It seems that 1N34 is now a catch all for anything Germanium. The best performers seem to be some diodes I scavenged from old computer boards I was given 30 years ago.

73 Mark VK6WV

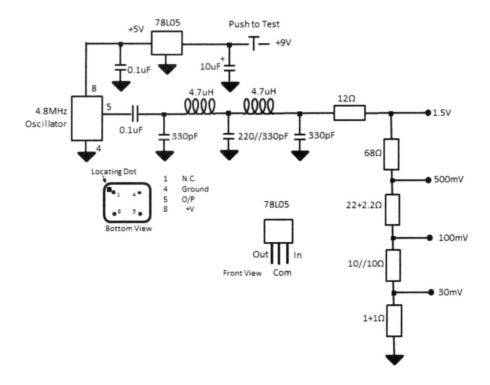

A REGULATED ADJUSTABLE HT POWER SUPPLY FROM 12 VOLTS

Chris Osborn G3XIZ   g3xiz@yahoo.co.uk

INTRODUCTION

Some of us are still interested in building and experimenting with valved equipment. Unfortunately mains HT transformers are quite rare these days and expensive, even on E-Bay
One way around is of course to put two mains to LT transformers back-to-back but this is not a very elegant solution and a fixed voltage HT is of limited use for experimentation purposes.

An adjustable and regulated QRP HT power supply, using easily available components is described here.

The unit is not powered from the mains but from a 12 V DC supply so may be of use for /P operation.
An ex-battery charger transformer was used 'back to front' but I have tested the circuit using a selection of mains to LT transformers and they worked well.
Needless to say small, low current transformers will give less available power.

The output may be varied from around 80 – 250 V and is useful for loads up to 6-7 watts.
Regulation is good with zero to full load not causing a volts drop greater than about 0.5 V
Full load ripple was in the order of 0.4 V p-p

PART 5 TEST EQUIPMENT

CIRCUIT DESCRIPTION

A 555 timer IC generates an audio frequency of about 300 Hz with a mark space ratio of 60%
The frequency may be adjusted over a limited range to minimize any transformer buzz and to optimize efficiency.

The HT output voltage is sampled via the potential divider R4 / R5 and compared to the set reference voltage on the inverting input of IC2. R4 may need slight adjustment on test to give the desired (or alternative) output voltage range.

When the sensed HT output voltage is lower than the reference voltage wider pulses are applied to the FET's gate via IC3. Conversely a higher HT output will cause narrower pulses and in this way is regulation is achieved.

The transformer secondary is rectified using a voltage doubler circuit and a simple R/C filter R11 / C10 removes the 300 Hz switching transients.
A meter is incorporated to display HT voltage and current and is of course optional.

An RF choke and decoupling capacitor may be inserted on the 12 V supply line to reduce any voltage spikes leaving the unit. I did not however find that this caused a problem.
R10 and C7 ensure that transients on the supply rail don't upset the pulse width generating circuitry.

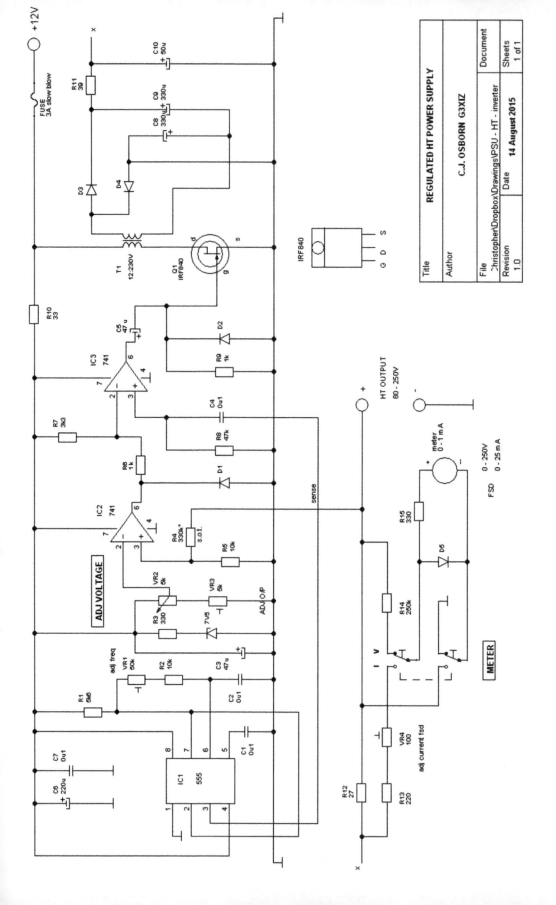

PART 5 TEST EQUIPMENT

CONSTRUCTION

An aluminium box was used to house the components (see photo)
The main circuit and metering components are mounted on veroboard and the IC's in DIL sockets.
Soldered header pins and sockets facilitated connections between the boards and the larger components.
High quality 4 mm binding posts were used for both input and output connections, the input being at the rear.
A 1 mA FSD moving coil meter gives switched 0-250 V and 0-25 mA indications.

CAUTION

As with all high voltage circuits care needs to be taken.
A sturdy metal enclosure with insulated terminal connections is recommended for safety and screening purposes.
The sensing circuit should bleed the smoothing capacitors to a low voltage within a few minutes.
Ensure that the fuse is a slow-blow type as the start-up current is considerable.

CONCLUSION

I built the unit three times with a slight variation of components and transformers and the design seemed quite robust.
A dual op-amp may be used in place of separate 741's
The other components may be 'tweaked' for optimum output and efficiency.

This a most useful addition to the shack's test equipment and has been handy for testing zener diodes, voltage regulators and varistors and for 're-forming' antique electrolytic capacitors.
It supplied HT to my home brew valve RX without any problems.

N CHANNEL JFET TESTER
ALAN TROY G4KRN alantroy49@gmail.com

There are different methods for testing bipolar transistors but testing common N channel JFETS like the 2N3819 seems to be more complicated. Here is an oscillator circuit with diodes in the output for reading RF voltage output on the DC range of a digital multimeter that we can use to test JFETS. It will demonstrate that the JFET oscillates and the output can be compared with similar JFETs.

The oscillator circuit uses a moulded 10uH inductor and oscillates around 14.380 MHz. The diodes are 1N4148s and allow a meter on the 20 Volt DC range to read the RF volts output when a JFET is inserted in circuit. The capacitor between drain and source is really critical and needs to be 10 to 12pF. I used two 22pF capacitors in series. A small trimmer of 33pF could be used and adjusted for best output.

I made up the unit on a piece of trackside veroboard with short, colour coded leads, for source, gate and drain to a small breadboard for inserting the JFET under test. Output leads go the the multimeter. Also leads go to a 9V battery. To operate, insert the JFET under test in the breadboard, noting the pin-outs, connect the multimeter on the 20V DC range and connect the battery and note the voltage reading.

Testing N channel JFETs from my junkbox: 2N3819s gave outputs ranging from 4 to 5V, J304s and J310s were around 5V and 2N5457s, an AF JFET were 2 to 3V.

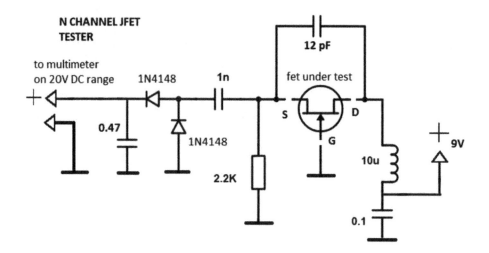

PART 5 TEST EQUIPMENT

SIMPLE SMART CONTINUITY TESTER
Peter Howard G4UMB, 63 West Bradford Rd. Waddington, Lancs. BB7 3JD

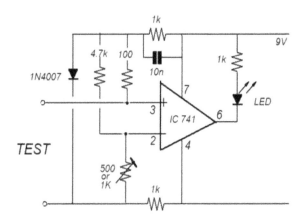

I have simplified a circuit that was published 40 years ago in Radio Constructor Magazine September 1977 by N. R. Wilson. It is a Smart Continuity Tester based on a Wheatstone Bridge which can be pre-set to indicate when a resistance is lower than a set value. Handy when tracing a wiring harness or where wires are connected through low resistance components. It can also be used to check whether earth paths are good. The setup is done by adjusting the 500 ohm variable resistor (I have used a multiturn type) to light up the LED at your selected test resistance. I set up mine up by connecting a resistance of 0.8 ohms across the test probes without the LED lighting up. But a dead short of the probes lights up the LED. The circuit wants a stable supply voltage and can also be run at 12v. At 9v the circuit draws 5 to 10mA. I have used a bright white LED from an old torch and tried different ones and all were OK. In addition you can add a piezo sounder across the LED to provide an audible indicator.

QRP SCRAPBOOK

Simple Electrolytic Capacitor Tester
Peter Howard G4UMB 63 West Bradford Rd Waddington Lancs BB7 3JD

It's unusual to find a circuit of a Simple Electrolytic Capacitor tester without using a signal generator with an oscilloscope to measure its series resistance So here is a simplified circuit which I made based on an article by G A French from a February 1977 Radio Constructor magazine .

You can use a multimeter switched to 50uA or 100uA range instead of buying a Meter. The meter readings are not linear so if you want to avoid having to make a calibrated scale for the meter or if you are using an external multimeter with a 1-100uA scale you will need to make a list of the uA values on the meter scale that correspond to various Capacitor values. So a batch of good assorted capacitors are necessary for calibration.

The original circuit had three ranges but this requires a three way rotary switch for S2. The third range switched in a 1000uF Cap. so that Range 3 went from 200 - 5000uF. The switches I used were all miniature 2 pole 2 way toggles. The meter needs to be set up only once. This is done as follows. Connect together the test points P1 & P2. Turn SW1 to short across the capacitors. Then adjust 470Ω variable resistor to read 0 on the meter, Then change the position of SW1 and adjust 100kΩ variable resistor so the meter reads FSD. Then remove the connection between P1& P2: Now the meter is ready to use:

To test. Put SW1 back to short across the capacitors. Put SW 2 so that it's connected to the 100uF capacitor and place another 100uF capacitor across the test points P1 & P2. Then change over SW1 to measure the CX. It should make meter needle read 50uA or half scale. This is because two 100uF caps are in series across the supply and the circuit is measuring the centre connection. I have found this unit to be reasonably accurate considering its simplicity and it will detect a leaky capacitor because the meter needle will keep moving up the scale. It will also act like a bridge circuit and can match two capacitors of the same value. Finally remember that every time you use it. SW1 has to short out the Capacitors first and then be switched over to measure CX. So I have labelled SW1 Measure & Discharge. My built meter has only one range so far, 20 – 500uF.

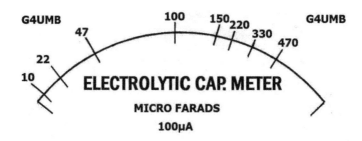

PART 5 TEST EQUIPMENT

ELECTROLYTIC CAPACITOR TESTER

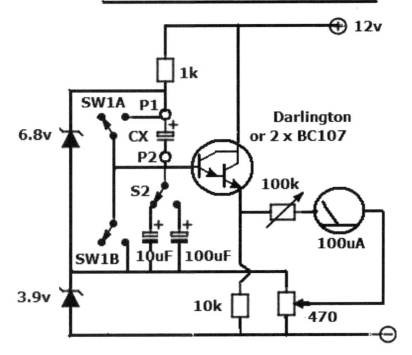

Range 1: 2 - 50uF
Range 2: 20 - 500uF

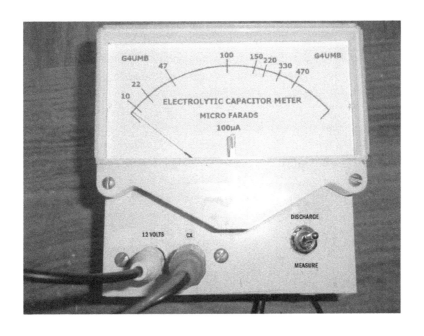

Junk-Box Valve Tester
Colin McEwen, G3VKQ colin@the-mcewens.co.uk

I acquired a quantity of 1940s vintage octal valves, mostly receiving types, nearly all 12-volt heaters, and wanted to find out how many of them were still usable.

I had enough components in my junk box [+ some standard stock items such as 1N4007 diodes] to build a valve tester. This led to some compromises on test conditions but avoided the risk of spending more on the tester than the value of the valves.

The circuit diagram is shown in the figure. The screen voltage supply of 85 volts was defined by the highest voltage Zener diodes in my junk box. The HT on-off switch and the large capacitor in the grid-bias supply are necessary to avoid a problem at switch-off. The design must ensure that the grid bias supply stays on as long as possible, certainly longer than the anode supply, at switch-off. If not, the valve may be overloaded when the grid bias drops to zero and the anode supply capacitors still have significant charge. The valves I wanted to test were 12V heaters, but transformers are available that give 6V if that is needed.

I dealt with the various pin-outs by connecting the 8 pins of the valve base, plus a flying lead for top-cap connection, to a 10-way "choc-bloc" terminal strip. I used another 10-way "choc-bloc" to provide feeds for anode [3x, see later] , grids 1, 2, & 3, cathode [3x], heaters, and shield [metal octal valves]. Cross connections between the two choc-blocs provided any combination of connections. More choc-blocs were used to connect the heater supply and two multimeters to measure anode current and grid bias.

Cross-connections were limited to 1:1 straps by providing 3 anode connections [anode, grid 2, grid 3, for triode-connected pentodes] and 3 cathode connections [for cathode, grid 3, shield].

I did not have a genuine top-cap connector but discovered that the top-cap is slightly larger than a BS1363 plug-top fuse, and therefore the Live terminal clip from a broken 13 amp plug could be used - with some careful bending out of the clip contact.

Test Method: Valve data sheets are readily available on the Internet. I set up the test voltages and grid bias specified in the relevant data sheet and measured the anode current. I found that quite a few valves gave at least 80% of the specified current with the worst being 50%. I stuck a label on each valve giving the %age figure as simple documentation.

N.B. Please be careful – lots of places to touch that could kill you!!

PART 5 TEST EQUIPMENT

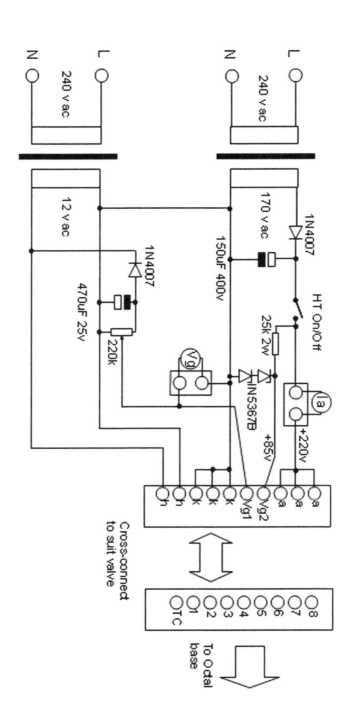

6 Miscellaneous

The Ultimate Bipolar V.F.O.
Gerard Kelly, G4FQN, 15 Dartmouth Drive, Windle, St. Helens. WA10 6BP
gerardkelly429@googlemail.com

Ceramic devices have been successfully used in the v.c.o. stage/s of many superhet transceiver projects for some time, unfortunately, they are not available in an extensive range of frequencies. I have a circuit for a 7MHz direct conversion transceiver based on an F.E.T. oscillator and thought a bipolar replacement must be as frequency stable and I thought better suited. A review of many articles from different sources showed the colpitts circuit to be the most common and a circuit was constructed. I must have built it using the wrong components because this circuit drifted, so I built another. Surely all these articles must have used better performing circuits than this? In the end I built several circuits and found that they all drifted. Almost at the point of giving-up and after a complete re-think the final circuit was under test and from the first switch-on I knew I had a winner. Without adding a P.L.L. or putting it in an oven, I cannot see any cost effective way in which it could be made more stable! Since then a circuit has been running at 2.6MHz in a superhet receiver (narrow band) without issue and tested over long periods at 7MHz; comparisons with an F.E.T. have not been made as this will have to wait for some more free time in the future.

The circuit is multi-purpose as it provides sufficient v.f.o. output for any transceiver or related project requirements making TR5 optional. As the operating conditions for TR1 are tightly controlled, component values do not need to be too strictly adhered to. Put your own favourite circuit in place of TR3. All the circuits were constructed 'ugly-style' on the copper side of a printed circuit board. Coils can be wound with three lengths of fine wire first twisted together in a hand drill.

Notes:
R1   A current-limiting resistor which controls the collector supply current
R2   Forms a negative-feedback loop stabilising the base biasing current
R3   Value chosen to limit the oscillator amplitude
R4   The 1k-1k5 / 100nF are extra decoupling to prevent r.f. feedback at higher frequencies – with very unusual effects on the carrier if it occurs
TR1-TR5   General-purpose n.p.n. A TR1 collector voltage of 3V8 appears to be the optimum.
L1   All coils were wound on a suitable toroidal iron cores
RFC   10 turns on a T43-50 Ferrite core or equivalent.

I am going to put the problems I had with this circuit down to - having all the right components, but not necessarily in the right order!!!

Seriously, if you are building a bipolar vfo do use this modified version and save yourself a lot of time.

PART 6 MISCELLANEOUS

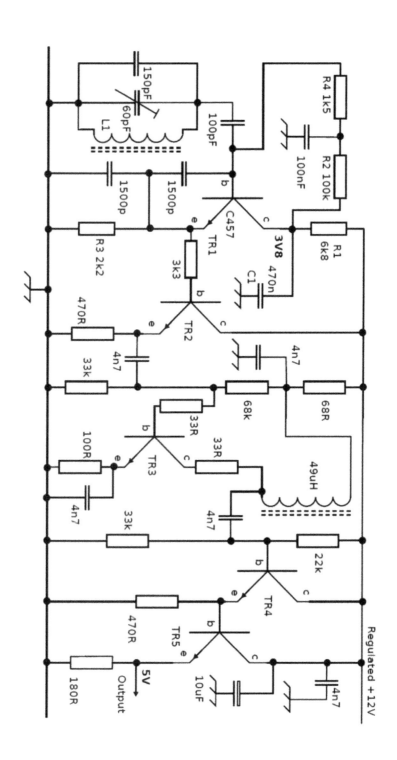

QRP SCRAPBOOK

> ### A Simple QRP 4m Transverter
> Roger Lapthorn, G3XBM, 37 Spring Close, Burwell, Cambridge, CB25 0HF
> http://www.g3xbm.co.uk   http://g3xbm-qrp.blogspot.com/

My homebrew rigs and FT817 allowed me to cover from VLF to 70cms but I had no way of getting on 4m (70MHz). With increasing activity and more countries appearing on the band each year there was a need to rectify this situation, so this transverter was the answer. The RX part is based on a circuit from a recent SPRAT and the TX part is an adaptation of a transverter design that I did some years ago for 6m. Several people have now copied the design and used it to get on 4m. Please feel free to do the same, but be prepared to optimise for your components and layout. Mine was done "dead bug" on a piece of copper laminate.

For the receiver, the NE602 is suitable as an RF preamp and mixer. These are cheap and pack a lot in a small DIL8 package. At switch-on it *immediately* worked copying the GB3BAA Tring beacon 89km away using my 10m halo as the (inefficient) antenna. Sensitivity on my generator looked fine. There is some out-of-band breakthrough as a result of the minimal selectivity in the front end and this may need to be addressed later with better input filtering, but has not been an operational issue yet.

For the transmitter side I used an SBL1 passive double balanced mixer but any similar DBM would be suitable. This is followed by a BF199 pre-driver (probably a 2N3904 will do here), a ZTX327 driver (a small Ferranti/Zetex device ex PMR) and an MRF237 TO39 RF PA (also ex-PMR). Similar RF devices should be fine. The TX strip was aligned initially using a 70MHz drive signal from a signal generator with trimmers between each stage. When optimised, some of the trimmers were measured with an LC meter and replaced by fixed value caps. This could also be done in the PA output PI filter too. Output is about 1W pep but with optimisation 2W pep should be possible. Adding the TX mixer and driving from the 28MHz source the power was the same with just a minor change of position of the tuned circuit in the pre-driver output. The TX strip is both stable and linear although the bias resistor values may need optimising. Keep the diodes used to bias the driver and PA in thermal contact with their respective transistors. A relay is used to switch RF paths between RX and TX and this is activated by a DC 5V available at the FT817 after a simple mod to the rig. This mod requires the addition of one wire

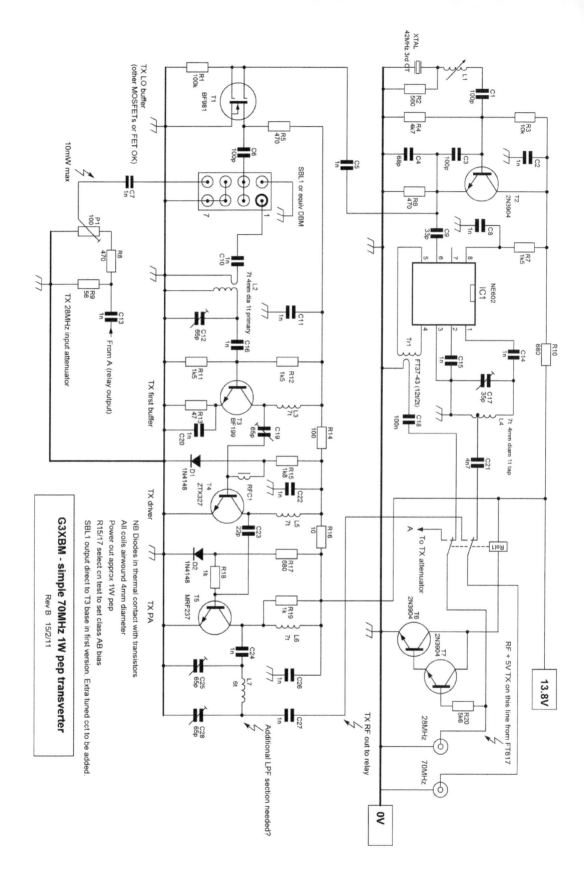

QRP SCRAPBOOK

from the TX 5V line, a 10nF capacitor and a 10k resistor. It is described visually on this site http://www.bergtag.de/download/ft817.pdf. Alternatively one could use an RF sensed TX-RX switch.

Several people have asked me about L1. L1 was a small 2 pin 4mm diameter coil with around 10turns of 0.2mm wire with an F29 core. It came from a PMR portable made by Philips. Its purpose is to ensure the crystal oscillates at 3rd harmonic on the right frequency. A technique that can be used is to remove the crystal (short it out) and make an LC tuned circuit (L1 and C1) that makes the "free running" oscillator oscillate at 42MHz. Then reconnect the crystal back in circuit and adjust either the L or C to bring back to 42MHz. It should be very close.

There are certainly improvements that can be made to this simple circuit, so use this as a starting point and experiment.

The main use of the transverter is to work European stations by summer time sporadic-E. For such propagation, a simple vertical omni-directional antenna is ideal, although if you intend more serious inter-G operation a beam like an HB9CV or a Moxon 2-el would be a better choice. With just a simple wire dipole in the loft my best inter-G DX so far is G4RFR at 229km on CW.

My website 4m page at http://sites.google.com/site/g3xbmqrp/Home/4m_transverter gives more information and has some videos of versions made by Mark MI0BDZ.

RF 20 dB AMPLIFIER FOR DC RECEIVERS
ERMITA #233 APTO 20, C. HABANA 10600 CUBA
cm2ir@frcuba.co.cu

I built this RF Amplifier for my Direct Conversion Receiver. I use a Transistor BF199 because it works very well for 7 MHz. (Ft max= 300 MHz)
I probe this circuit in a HAMEQ Tracking & Spectrum Analyser, and the frequency response graphic was 20dB from 3 MHZ to 35 MHz with very good linear performance! You can regulate the amplification with a potentiometer replacing the 300 K resistor.

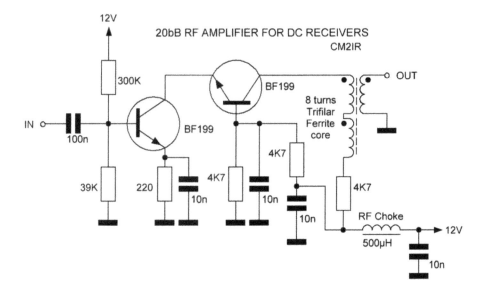

Fig. 1 Diagram Schematic.
All capacitors are Polyester or Silvered Mica. The resistors must not be inductive types. The Balun can be 4:1 or 6:1 because output impedance of an RF Cascode Amplifier design is around 200 Ω to 300 Ω. The input capacitor is 0.1 μF with a resistor to ground of 39 KΩ. Input Impedance: 50 Ω. Output Impedance: 50 Ω

In the balun I use a little ferrite core but I haven't a serial number or type, but I think that you can use any ferrite because the frequency is low (7 MHz). I used three twisted wires (0.12 mm) with 8 turns around the ferrite core. If you have a Spectrum Analyzer and Tracking you build two baluns and connect as 50Ω - 50Ω (IN-OUT) in each one you can observe the attenuation graphic of ferrite core.

QRP SCRAPBOOK

Make Your Own Ribbon Cable
Anthony Langton, GM4HTU, 71 Gray Street, Aberdeen AB10 6JD

I was building a piece of audio test equipment, made up of seven modules connected together by Molex style pcb headers and housings. These are ideally suited to ribbon cable, which makes a very neat job, rapidly assembled. Unfortunately, the only ribbon cable I have is ex PC, all grey apart from one red tracer. I like to use colour coded cables, especially on a project with so many connections

I went for the traditional SPRAT reader's solution: make my own. One of the problems with plastics is that each type seems to need its own glue. A friend who had just finished a plumbing job suggested PVC pipe joining glue would probably stick PVC insulation, and loaned me a part -used pot of Flo-Plast Solvent Cement. It is a clear, slightly viscous liquid with the most appalling smell. The instructions with the cement advise working in a well ventilated place. This should be heeded. I made up a simple fixture to hold two pieces of insulated wire in close proximity and applied the cement between them, using the brush provided in the lid. It was not a precision job: it unavoidably went everywhere. I wiped the wires down with a paper towel, which cleaned off the surplus and had the added benefit of squeezing the wires together. I tensioned the wires then went for some fresh air.

A while later I returned to the workshop to examine the results. The smell had subsided somewhat and the outcome was very much better than I expected. My fear that the excess glue would damage the insulation was unfounded; it had not even taken the shine off. The bond between the wires was strong and needed a good pull to separate them. Really well joined pieces needed cutting. Buoyed by this success, I bought my own pot of glue and tackled a 1m length of red, black and blue. This time I worked vertically, suspending one end from the workshop roof and weighting the other with a toolmaker's clamp and some scrap metal. The results were most satisfactory. For shorter lengths I refined the original test fixture so I could use it in the garden and avoid stinking out the workshop.

The project has been so successful I have given up lacing and twisting cables and now try to use ribbon cable for everything. The photo shows the front panel assembly of a sweep oscillator. The wires do not have to be side by side: you could assemble two on two as a quad, five in the shape of a quincunx or whatever you require. Have a go, it's quick and easy and makes a really neat job. Just make sure you have good ventilation.

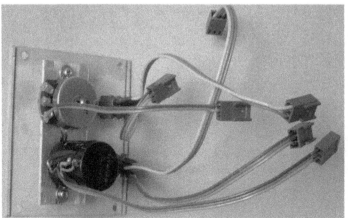

PART 6 MISCELLANEOUS

Occam's Microcontroller
Paul Darlington - m0xpd - 8 Uplands Rd, Flixton, Manchester

It might be supposed that the internal complexity of microcontrollers, digital synthesizers and similar devices places them somewhat at odds with our guiding principal of *"economy of means allied with richness of result"* [1]. It is, however, the purpose of this article to argue that recent developments in microcontroller technology and the ready availability of other digital subsystems have made these resources an efficient foundation for the development of minimalist radios. Further, these developments have been such that potential benefits are accessible to all experimenters, rather than just those who have special interest or expertise in digital methods.

Microcontrollers are "small computers on a single integrated circuit containing a processor core, memory, and programmable input/output peripherals". They are intended for *embedding* into larger systems, where they serve as controllers and communication channels. They have been used in peripherals of the rigs featured in these pages for several years, serving in roles such as keyers, frequency displays, power & SWR displays, etc. However, this article will explore by example the idea of embedding the microcontroller right into the heart of a QRP rig, rather than being content to leave it on the periphery.

To date, two important obstacles have limited penetration of embedded computing into the working vocabulary of home-brew QRP enthusiasts; complexity (real or perceived) and inflexibility (especially those designs presented in "closed-source" format). We have seen laudable attempts to demystify microcontrollers, including in these pages [2], but those initiatives have not yet overcome the barriers to widespread adoption.

Winds of Change
Fortunately, a quiet revolution has been taking place in the world of microcontrollers, embedded systems and "physical computing". This revolution was not targeted at technically sophisticated beneficiaries – rather it was intended for artists, hobbyists and schoolchildren. Radio amateurs, with their explicit technical skills proven by licensing, are well equipped to thrive in this post-revolutionary "new world".

The quiet revolution has not been achieved through change in the microcontroller devices themselves – they still express the same internal architecture and still offer the same features as before, albeit at continually falling cost/performance ratio. Rather, the devices have been presented in different contexts, designed to make it easy for users to exploit their power through accessible, intuitive programming languages, easy to use (and usually free) development environments and powerful, flexible, inexpensive and standardized hardware platforms. All of these make it easy to get a microcontroller to do a useful job of work – like form the backbone of a QRP rig.

The opening steps in this revolution have been played out in full sight of G-QPR members, who saw the original PIC microcontroller [2] "re-cast" as the "PIC-AXE", which was featured at a recent G-QRP mini-convention [3] and is available through club sales. The

QRP SCRAPBOOK

PIC-AXE *is* a PIC - but a PIC that has been tamed by the provision of new, simpler means to interact with it and exploit its features. There is a clearer (*and much more successful*) embodiment of this revolution in the world of the AVR brand of microcontrollers, in which the ATMega device has been "re-cast" as the "Arduino".

Arduino is more than a subset of the AVR microcontroller, underpinned by software resources. It also has physical expression as a family of boards, making it truly an "electronics prototyping platform". These boards bring the resources of the microcontroller into easy reach, making it laughably easy to connect to systems such as our radios – as we shall see. A board-level device called the "Pinguino" is an analogous system for the PIC microcontroller.

Most importantly of all, the revolution in microcontroller accessibility is open–source, both in software AND hardware. This makes it not just possible but actively encouraged to share and build on the efforts and experience of a vast community of users.

The argument of this article is agnostic to the particular brand and family of microcontroller used. However, we shall present all subsequent examples in the context of the Arduino platform.

Manna from Heaven

In addition to developments in the way microcontrollers are supported, described above, an interesting crumb has fallen from the table of high-tech electronics. Opportunistic QRP enthusiasts have been quick to pick up this crumb – which takes the shape of a "Direct Digital Synthesis" Signal Generator Module, today widely available for less than the price of the (AD9850) chip it contains. With this module it is possible to generate stable, controllable sinusoids "from dc to light" (or, at least, to beyond the top of the HF band). All that is required is some means to control the module and communicate with it – a perfect application for a microcontroller.

Commenting on his QRO power and SWR meter, [4], Ray, g4fon, observed "*that meters have become expensive, whereas an LCD display might ... actually be less expensive and more robust*", fully justifying the use of digital building blocks. Similarly, we should notice that a microcontroller and a DDS module is cheaper than a decent quality variable capacitor and reduction drive, such that a digital VFO is now not just a feasible technology but also a literal "economy of means". However, controlling either a SWR meter or a VFO will leave a microcontroller twiddling its thumbs for most of the time. It could undertake both tasks - and more...

PART 6 MISCELLANEOUS

If we assemble a partial list of the features within a practical HF CW transceiver that could be managed by a microcontroller, the possibilities are impressive:

Keying	**Display**
Iambic Keyer	Frequency
Automated "CQ" calls	SWR/Power
Beacon Operation	Field Strength
Frequency Input	
from dial / keyboard / memory	**Tx/Rx Switching**
	Tx Keying
Frequency Control	Rx Muting
DDS Control	QSK / Semi break-in delay
Tx/Rx Offset and "RIT"	

Notice that some items of this list are possible only because of the use of a digital device (*management of a DDS chip or module for the local oscillator, digital frequency input and display, etc*). Other items of the list might previously have been interpreted as functionality normally residing outside the rig (*such as the keyer*). Use of the microcontroller allows us to extend the scope and ambition of our homebrewed rigs and make them more complete as self-contained communication systems - like commercial rigs. However, there are items on our list that are "logical" (i.e. essentially DIGITAL) tasks, which would have been required in the rig whatever technology was used. This is particularly seen in the management of transmit and receive switching. QRP rigs have always included such switching functions and their implementation in a microcontroller will be found to add considerable flexibility.

Obviously, we could extend the list to add any number of "higher" functions (such as automatic adjustment of an ATU) – but we shall deliberately stay well clear of the slippery slope that leads towards software defined radios and similar complexity. Our interpretation of "economy of means" requires that we "keep it simple".

Example: A QRP CW Transceiver
We now present an example of a practical and successful CW rig, built on the principles outlined above. The rig has made CW QSOs on the 80, 40 and 30 metre bands and has operated as a beacon in QRSS and various FSK modes. It is not the purpose of this article to be proscriptive, so the details of the rig are not important. Instead, we shall focus only on those points of connection and interaction between the microcontroller and the rig. The building blocks of the rig are typical of ordinary QRP practice and so these "points of connection" may provide inspiration for readers to modify their own favourite circuits for micro-control.

The overall architecture of the example rig is seen in the figure over the centre pages, which includes details of the interface between the rig's sub-systems and the microcontroller, discussed below.

QRP SCRAPBOOK

Keyer

It is useful to start with the iambic keyer – for two practical reasons. Firstly, a keyer is a very useful "first project" for those new to microcontrollers, expanding on the traditional "press a button and light an LED" application, common to most microcontroller training courses. Secondly, the keyer actually will form the "backbone" of the software for our example rig.

The keyer's hardware is trivial. One side of each switch of the key and/or paddle is connected to an input pin of the microcontroller, with the other side of the switches commoned to ground. Most microcontrollers feature internal pull-up resistors, which can be enabled on certain inputs. These pull-ups hold the input "high" until the key is pressed, which action will short the input down to the alternative "low" logic state. The key's state is sensed by sampling the voltage on the input pin repeatedly in the main operating loop of the software and watching for a change between input voltages on two successive observations. A change from "high" to "low" voltage indicates a key press and the opposite transition indicates a key release. The time taken between observations of the input (time during which the microcontroller is attending to other tasks) creates a switch de-bouncing delay.

It has been found possible to build not only the keyer but also the entire transceiver without using the "interrupt" facilities of a micro-controller, thus keeping it simple. The resulting rig does not have any noticeable latency between operating the key or paddle and hearing the sidetone (which itself is generated by a native function of the Arduino language).

The same code which implements an iambic keyer can easily be expanded to read a message saved as text, one character at a time, look up the Morse equivalent of each character and send it at a pre-determined speed, making possible automated CQ calls and beacon operation.

Local Oscillator

The combination of a DDS chip (or module) and a microcontroller make possible a flexible RF generator with range, stability and accuracy that matches or surpasses conventional analog oscillators traditionally used in our rigs. We shall examine the hard- and soft-ware required to make a flexible VFO in a moment but, for now, we shall consider what is required just to set the local oscillator running at one frequency. The DDS chip typically will expect to receive configuration information from a controlling device in 'serial' format. This is to be preferred over the 'parallel' alternative as the number of available pins on any microcontroller is limited.

In the case of the Silicon Labs' Si570 device (used in the excellent USB synthesizer [5] and pa0klt VFO [6]) translation between the desired frequency of operation and the configuration code required to produce this is not simple. However, the AD9850 device accepts a numerical input proportional to the desired frequency, making programming very easy...

PART 6 MISCELLANEOUS

Here's the Arduino code (developed from [7]) required to calculate the configuration data and send it to the AD9850...

```
// calculate and send frequency code to DDS Module...
void sendFrequency(double frequency) {           // 'frequency' is the desired f
  int32_t freq = frequency * 4294967296/125000000; // module has 125MHz xtal
  for (int b=0; b<4; b++, freq>>=8) {            // 'freq' has 4 bytes...
    shiftOut(DATA, W_CLK, LSBFIRST, freq & 0xFF); // sent one at a time
  }
  shiftOut(DATA, W_CLK, LSBFIRST, 0x00);         // final "0" control byte
  pulseHigh(FQ_UD);                              // and a pulse on FQ_UD
}
```

The DDS module hosting the AD9850 offers both a parallel and a serial interface – of course, we choose to use the serial interface, which corresponds to the pins named "DATA" and "W_CLK" in the code segment above. Other connections (for power and two control lines, FQ_UD and RESET) are shown in the main figure. All connections to the module are through an easy-to-use 0.1 inch header – which strongly differentiates this module from the rather less user-friendly physical interface to surface-mount DDS chips.

Keying the Tx
With a local oscillator generating a stable source of RF and means to generate the timed sequences of Morse code, all that is required to transmit CW is means for the microcontroller to key the transmitter. A single transistor, in open collector configuration, replacing the physical "key" in a conventional QRP transmitter design, easily achieves this. The Tx detail in the main Figure shows an example of this approach applied to an early stage of the "Ugly Weekender" [8] transmitter.

The building blocks described to this point - keyer, oscillator and means of keying the power stage - add up to a simple CW transmitter or a QRSS beacon. This transmitter or QRSS beacon is more flexible than the traditional rock-bound alternative, as it can QSY to any frequency with a simple re-flashing of the microcontroller' program. However, the full potential of the microcontroller-based rig only begins to be unleashed once we exploit some of the dynamics of the DDS module and add a receiver...

Varying the Oscillator
The DDS module reacts very quickly to a command to change frequency, with negligible disturbance of the output waveform at the instant of transition (indeed, the transition is phase-continuous). This, coupled with the speed of the microcontroller, makes the oscillator frequency-agile, facilitating FSK modes (such as WSPR [8]), the transmit/receive offset inherent in CW operation and frequency changes to tune across bands and switch between them.

The simplest means to control the frequency of a DDS local oscillator is to read the voltage on a potentiometer wiper, configured as a potential divider between Vcc and ground. Most microcontrollers feature integral analog-to-digital converters. These are capable of directly reading such a voltage to (typically) 10-bit resolution (*i.e. they are able to resolve to ~ 0.1% of full-scale-value*). Thus, with the addition of a single, simple potentiometer, we can make the previously rock-bound transmitter described above able to tune over useful segments of a single band. For example, a 5kHz section of the 40m band can be covered with tuning that sounds and feels continuous in "pitch". This is sufficient to cover both the QRP and FISTS frequencies of interest to many readers.

Such a simple means of adjusting the frequency of the local oscillator will be entirely familiar to QRP enthusiasts, who are used to VFOs with single-knob tuning from a variable capacitor (or potentiometer biasing a varactor diode). The knob provides not only the means to adjust frequency but also a visual indication of the frequency setting. If we wish to fully exploit the flexibility of the DDS module, we must be able to monitor the frequency setting with greater resolution. This can be achieved by a link from the rig to a PC, as the Arduino offers a pair of pins dedicated to such a link and the software tools to send the current frequency to a PC with a single line of code…

However, in requiring the presence of a PC, such an approach is neither minimalist nor always convenient. Fortunately, the microcontroller can easily drive a display…

Display
A numeric or alphanumeric display can provide useful information about the state of a rig, such as the power / SWR monitoring function already mentioned. However, we shall consider the display of frequency as the only necessary function and use it as sufficient example of the integration of a display into a micro-controlled rig. As Ray, g4fon, observed [4], LCD displays are robust and inexpensive and entirely suitable for monitoring the frequency of our DDS module's emissions. There is, however, one obstacle to their use.

Many alphanumeric displays are built using an interface associated with Hitachi, which has an 8-bit data bus and some further control lines to latch data into and out of the device. As already has been mentioned, microcontrollers have only a limited number of input/output pins and this Hitachi interface is expensive in these important resources. Fortunately, there are a number of means by which we can overcome this potential obstacle.

First, it is always possible to upgrade the choice of microcontroller or microcontroller system to increase the number of available I/O pins (e.g. changing from Arduino UNO to Arduino MEGA) – but this approach is the antithesis of minimalism! Second, it is possible to add simple shift registers to make serial-to-parallel converters, by which a small number of I/O pins (on the microcontroller side) may be connected to a large number of peripheral pins – but this approach only really starts to deliver benefits for a large numbers of pins. Third, we can use the Hitachi interface in an alternative "4-bit" mode, in which commands and data are transferred a "nibble" (i.e. half a byte) at a time, saving 4 pins. Fourth, we can use (or, better still, make) a serial-to-parallel converter module, which accepts a two-wire input, such as I2C, from two pins of the microcontroller and connects to the Hitachi interface. Such modules are now widely available for a few pounds.

A fifth approach was used in the development of the rig used as example in this article, which exploits a 12-character numeric LCD display with serial input, found in the junk box! The data line is actually shared with the data line to the DDS module, further reducing the load on microcontroller I/O pins.

Rotary Encoder for QSYs

The finite range of adjustment available from a potentiometer is radically expanded by use of a rotary encoder, which generates digital outputs on rotation of a familiar "knob". These digital outputs can be interpreted by the microcontroller to produce a very flexible means of frequency input, allowing adjustment over a wide range at controllable sweep rates or resolutions. Rotary encoders have two internal switches, which switch in quadrature as the shaft is rotated. The resulting two-bit sequence encodes direction and (if required) speed of rotation. The sequence can be read by very simple logic, as demonstrated by this Arduino code segment…

```
// (as part of the main loop structure…)
RotEncA = digitalRead(RotEncAPin);       // read the two rotary encoder pins…
RotEncB = digitalRead(RotEncBPin);

                                         // if there's a state change on input A…
if ((RotEncA == HIGH)&&(OldRotEncA == LOW)){
                                         // AND input B is LOW…
if (RotEncB == LOW) {
                                         // then knob was moved clockwise…
  Frequency = Frequency + df;}
else {
                                         // otherwise, moved anti-clockwise…
  Frequency = Frequency - df;}
}
OldRotEncA=RotEncA; // save the input A value for next time round the loop
```

We can change the frequency increment (*"df" in the code above*) to give an appropriate rate of frequency control; df=10Hz is small enough for "continuous pitch" tuning of CW, whilst df=1kHz gives a reasonably quick sweep across the entire band. Many rotary encoders feature a third internal switch, operated by a push on the knob. This can be used to toggle through a set of different values for the frequency increment, df. The example rig uses 10, 100 and 1000 Hz.

With frequency inputs passed from a rotary encoder to the DDS and monitored by an LCD display, the oscillator becomes a truly flexible VFO, which can tune the transmitter described above over multiple bands.

Transmit Muting and Break-In

The same logic signal used to key a CW transmitter can be used to simultaneously mute a receiver, making an entire transceiver. The interface between a microcontroller I/O line and a practical Rx mute is shown, in the example rig, in the context of the FET passgate used in the receiver of Roy Lewallen, w7el's "Optimised Transceiver" [9]. Although it is possible to share the Tx keying and Rx muting on a single I/O line, having these functions separated allows greater flexibility in managing break-in delays.

In the example rig supporting this article, keying the transmitter initiates a counter, which implements a fixed delay. The receiver is muted for the whole of this delay period. The same delay manages the frequency offset required between Tx and Rx. If the transmitter is

QRP SCRAPBOOK

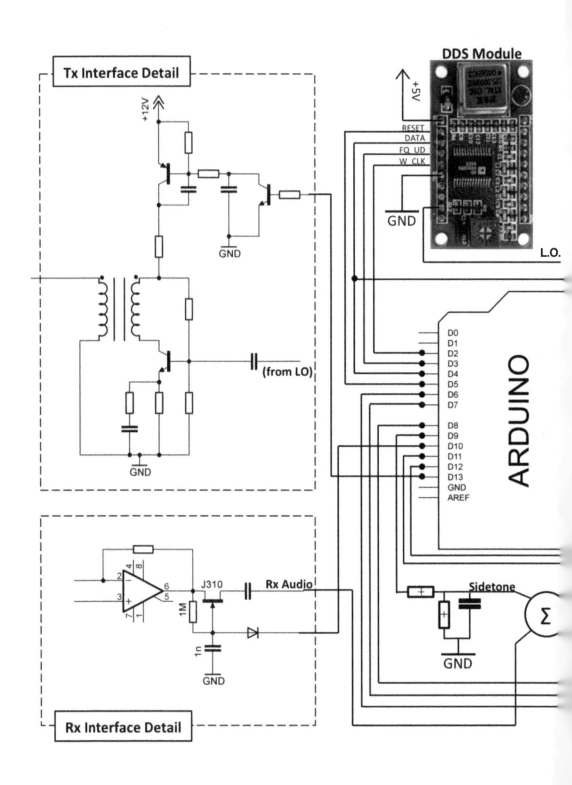

PART 6 MISCELLANEOUS

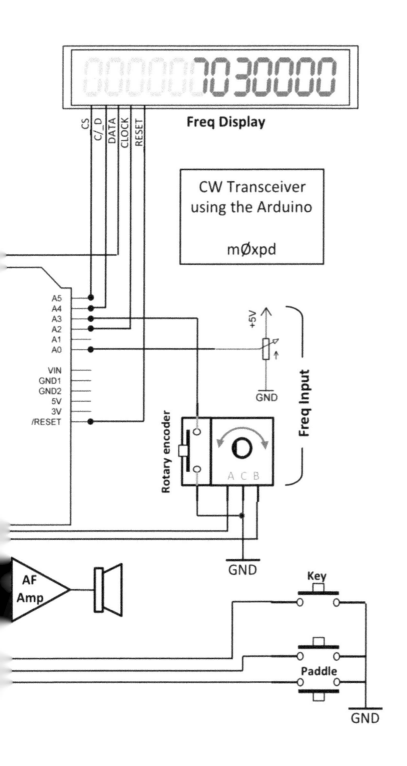

QRP SCRAPBOOK

keyed again during the delay interval, the count is re-started. The start value of the count determines the mute interval – a very small count value gives full break-in, whilst a larger value holds the receiver muted over individual Morse characters or words.

The potentiometer used for simple "analog" frequency input can be exploited as a "Receive Incremental Tuning" input to vary the transmit offset, where main frequency input is via a rotary encoder.

Code Samples, Schematics and a Note of Caution
Although the example rig supporting this article was intended to illustrate methods and inspire others to experiment with their own microcontroller-based QRP systems, there may be further details of its construction and code of interest to readers. A full schematic and a series of Arduino programs ("sketches") are available for download at: http://www.gqrp.com/sprat.htm

To date, only one downside of the example rig has emerged; the DDS module draws 200mA, which clearly limits its usefulness for /p operation.

Whilst many QRP enthusiasts enjoy using anachronistic or even obsolete technologies in their h/b rigs, the microcontrollers and other digital resources used in commercial transceivers are now available and accessible for widespread exploitation. This digital technology can offer "economy of means allied with richness of result".

Notes and References

[1] D L Sayers paraphrases Occam's Razor in these words in her novel "The Five Red Herrings", 1931
[2] "Starting with PIC", P Debono, 9h1fq, SPRAT 138, p17
[3] "A Maine Yankee in Rishworth Court", R Harper, w1rex, SPRAT 141, p34
[4] "QRP Wattmeter", R Goff, g4fon, downloaded from:
http://www.g4fon.net/wattmeter.htm
[5] QRP2000 USB Controlled Synthesizer Kit:
http://sdr-kits.net/QRP2000_Description.html
[6] pa0klt "Low Noise VFO Synthesized Kit":
http://sdr-kits.net/PA0KLT_Description.html
[7] Testing an eBay AD9850 DDS module with Arduino Uno, R Rollinson, nr8o:
http://nr8o.dhlpilotcentral.com/?p=83
[8] "Ugly Weekender", R Hayward, ka7exm & W Hayward, w7zoi, QST, August 1981 – see also Todd Gale, ve7bpo's notes at: http://www.qrp.pops.net/transmit.asp
[9] "An Optimised QRP Transceiver for 7 MHz", R Lewallen, w7el. QST, August 1980 – now available at:
http://www.arrl.org/files/file/Technology/tis/info/pdf/93hb3037.pdf

PART 6 MISCELLANEOUS

Universal SSB Generator
David Smith G4COE, 54 Warrington Rd. Leigh, Lancashire. WN7 3EB
davecoe@blueyonder.co.uk

Jandek produced a DSB generator kit that used 10.7Mhz (see Sprat issue 76 Autumn 93, page 25~26 for full details). This is a modified version for 9 Mhz, the SL6270 vogad chip done away with being far too expensive and hard to get, an op-amp vogad is used instead.

Rather than have resistors standing up on end I used the tinier 0.125 Watt types, they make a neater appearance but most importantly no long leads. Some eyes may frown not using double-sided pcb. I feel you get allsorts of added sporadic capacitances that does more harm than good – never ever build VFO's on them for sure except ugly construction style!

One catch, unless someone finds an alternative, there are three Toko coils you need to rewind, the hardest part of the whole project is getting the coils apart, I used KANK333R types with a red core. A direct replacement is the Spectrum Communications 45u0L coil available from Club Sales. There are only two used on the board, the third one being on the ssb filter board.

The whole lot, mic. vogad amplifier, MC1496 balanced mixer, crystal oscillators and the first fet amplifier that drives the balanced modulator all sits on a 4¼ x 2½ inch single sided pcb.

Because of varying filters sizes, the filter and the second fet amplifier was left off the board and its layout left to the constructor. This makes it more 'universal' and adaptable for a wider choice of filters and other filter frequencies. One has just to alter the crystals and the three transformers – no reason why ferrite cores shouldn't be used here, I prefer them being 'tuned' which means using a trimmer capacitor, the broadband types I feel isn't as good in this application, because tuning helps to keep our signal clean and making far better filters.

By building an off-board filter unit we aren't being limited, we could make the filter part of the receiver, a defunct commercial rig ought yield other SSB filter frequencies of varying sizes, we then only need to rebuild the filter board and retune the transformers.

With our 9Mhz signal coming out the filter board we can really go to town... mix this with a 5.0-5.5 Mhz VFO followed by a band pass filter for either 14 or 3.5 Mhz we have a SSB signal ready for amplifying. Nature dictates the 3.5Mhz band will tune backwards in this case.

To solve this 'backward tuning' we could build a mixer vfo, not only will this solve our problem it'll enable us to add other bands as well. We could go 'all mod cons' and use a synth vfo module.... who said building an ssb transmitter was difficult?

Building a ssb transceiver sounds like one heck of a task, do it in stages and it's like shelling peas, start off with the basic transmitter then add a receiver. For cw you could key a 9 Mhz crystal oscillator insert it before or after the ssb filter, another way is to key a audio 'ossy' into the mic socket or audio input of the balanced modulator itself depending on it's level.

Only two things to worry about, well three really, setting the signal level, vfo stability and the third....... not ending up in casualty getting them 'Sonnuffa guns' apart to rewind!

QRP SCRAPBOOK

Winding the transformers:
Two T1's and one T2 Toko 10K KANK3333R (or Spectrum Communications 45u0L) are required, the hardest part is getting them apart… no reason why you can't get loose bobbins, cut the casing off one to obtain the red coloured cup but it means having some casings handy, the red cup must not be damaged in anyway. We need to remove all the wire on the former before we can start rewinding, in the component section a link is given to obtain an excellent article on rewinding Toko's and how to 'get at them' by G4HUP in pdf format. The wire used must be around 40~38SWG, these can normally be had from relays and transformers. All windings should be done in a clockwise motion. OK, let's identify the pins-:

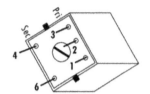

Primary:
Pin 1 Start
Pin 2 Tap
Pin 3 Finish

Secondary:
Pin 4 Start
Pin 6 Finish

T1 and T3: Primary 1 turn and secondary 16 turns.
With some 40swg e.c.w, bare and tin the ends secure to pin 4 by a single loop and solder, feed the wire to the bottom pile and wind 1 turn feeding this to pin 6, again bare and tin apply one loop round the pin and solder. Secondary windings are always wound first.

Now for the secondary, again secure the wire but to pin 1 this time, feed the wire to the second pile apply 8 turns then feed it to the next pile and wind another 8 turns feeding it out to pin 3, bare, tin and secure as before then solder – that's' it!

At this stage I temporarily soldered an 82pF capacitor across pins 1 & 3 and checked with a gdo, no need to put it in its case at this stage but do put the red cup on top! Dipping it at 9Mhz, if all is well put it back into its case. To identify them I marked them with a spot of paint, remembering these two coils the same.

For T2: Primary 8+8 turns and secondary 2 turns.
Wind as for T1 but using 2 turns on the secondary, the only difference is the extra turn.

The secondary is the same but requiring a centre tap. Starting at pin 1 wind 8 turns in the next pile then lead this to pin two – the centre pin, by being very careful we can bare and tin a portion where it loops round pin 2 in a figure eight style to allow us to solder, far better than using two separate lengths of wire, after our first 8 turns have been securely attached to pin 2 we can continue, feeding and winding another 8 turns in the third pile, lead out, tin and secure to pin 3. Just like we did with the other transformers we can now check this with a gdo and an 82pF capacitor after popping a 'Red top' on before sliding the casing on…. That's it!

Important: When winding ensure the wire goes securely into the correct pile and hasn't snagged or jammed on the sides and that there isn't any slack or kinks in the wire, keeping it nice and taut. Ensure the 'Red top' is pushed down firmly before checking and fitting the case. No wax or glue is required anywhere to aid fixing. Pin 2 is only used on T2 for the MC1496.

The tuning range was just beyond 10Mhz and down to about 8Mhz on all coils I wound checking with a gdo, poking the gdo coil through the capacitor leads was enough.

38swg and 40swg wire did the trick admirably. To give an idea of the length of wire needed, 18 inches of wire for one secondary left enough for the ends, have a 'practice wind' first if this is your first attempt… no need to solder the ends though.

PART 6 MISCELLANEOUS

The alignment.
First of all with no power, check you have the highest resistance with a test meter between R3 end nearest PR1 and ground by adjusting PR1, turning it the opposite direction will reduce the resistance, this point is common to all the oscillators and gives maximum rf output.

Apply power to the relevant crystal, not forgetting the '+12V always' terminal must be powered as well, check the oscillators are running on a receiver, then with a rf probe or oscilloscope connected to CIO out & TP (test point) peak T1 for maximum output with the 9Mhz crystal. Applying a frequency counter to our TP will now allow us to set the correct crystal frequencies with TC1, 2 or 3 and should be set for CW - 9.0000Mhz, LSB - 8.9985Mhz and USB - 9.0015Mhz, the crystals can be fitted in any order to suit the builder's preference.

The next step is to adjust the drive level at TP with a oscilloscope, RF voltmeter or on a rcvr, adjust PR1 for 300mV RMS or 850mVp-p appx. this may need a 'tweak' should you use the output for the product detector in a transceiver configuration, this point also feeds the balanced modulator via C34 as correct drive level is required for proper operation.

Setting Carrier balance, to begin with ensure PR3 is set to one side of its travel, any side will do. The +12V TX terminal now needs to be powered leaving the other two powered up as well. Now connect an rf voltmeter or oscilloscope to the dsb output terminal near T2 and peak T2 for maximum output with a crystal selected, after T2 has been peaked adjust PR3 for minimum output, this affects the carrier suppression.

Finally, set the microphone level by adjusting PR2 for 71mV rms or 100mVp on pin 1 of IC2. We can apply a 1.5Khz signal to pin 1 of IC2 or a low level signal to the mic. skt.

Condenser microphones require a small dc bias, this is achieved by linking the bias pin to the mic. input pin on the pcb. The overall gain can be controlled somewhat by varying R12 1M ohm resistor, this will vary between microphones. Since R19 is in series with PR2 this can also be used to 'tailor' the audio level to the balanced modulator. We can monitor our dsb signal on a receiver, by hooking a length of wire as an antenna to the dsb output terminal as an antenna.

Now that we have our dsb signal the builder needs to build a filter board for the filter being used. The ssb filter will feed the second fet amplifier, being like Q4 and T3 as the tank. T2 and T3 cater for filters with a 500 Ohm termination. We don't have to use a ssb filter we could transmit dsb but does mean half the transmitted power is wasted through the unwanted sideband - an awful waste!

Bits n' bobs.
Three pcb's were made successfully with 'Press and Peel' toner transfer. A link to ground is made on pin 14 of IC2, this can be on either side of the board to aid earthing. J8 link is obviously not required if using a metal screen, I used 0.5mm brass for the screens soldered to solder pins, to aid soldering I applied a tiny bit of solder flux. There 'could be' two problem areas on the pcb caused by 'close passes', this is around the secondary pins of T2 and the tracks between pins 4 & 5 and pins 11 & 12 of IC2. There were no problems resulted when soldered but they are close!

Playing with the crystal oscillators I discovered those polypropylene trimmers were 'yak' regarding frequency stability, same with those mini. Philips ceramic plate capacitors, so I used the NPO disc types and ceramic trimmers here and the stability was far better than those 'poly wobblers' that produced a curious cyclic up and down drift cycle on the frequency counter, we're talking hertz here!

The 'fat resistor looking' 150uH rf chokes aren't critical but don't go below 100uH, finally pins 3,7 and 9 of IC2 MC1496 are not used they're blanks!

QRP SCRAPBOOK

PCB tracking. <u>Viewed from component side.</u>

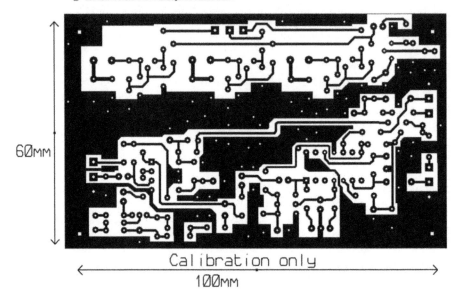

Vogad circuit.
This can be built as a stand a lone unit for other uses and powered by a battery. R19 originally was 150K along with PR2 being 1K Ohm no doubt ideal for feeding into a 600 Ohm mic. input socket. In our case R19 is reduced to 1K and PR2 changed to 5k.

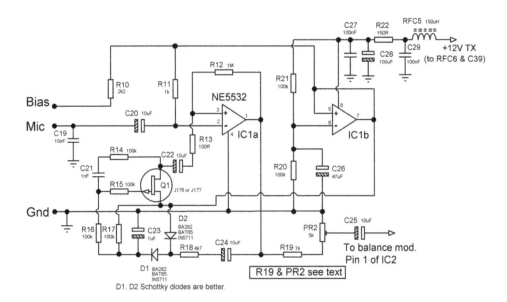

PART 6 MISCELLANEOUS

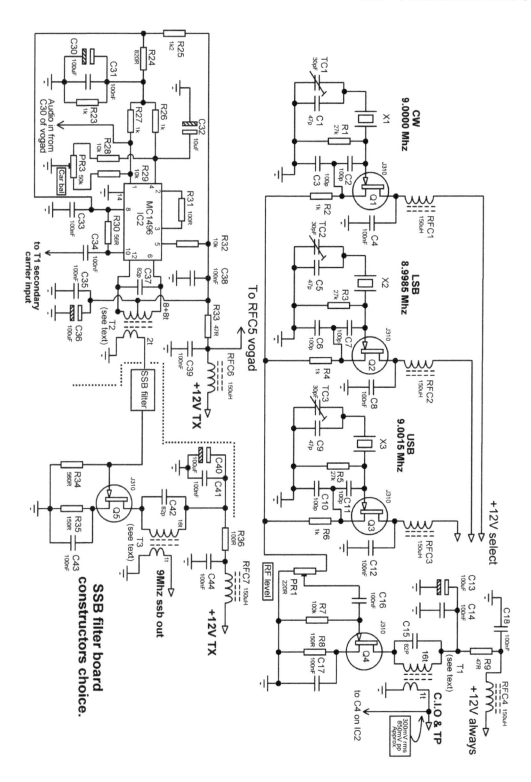

Component Layout.

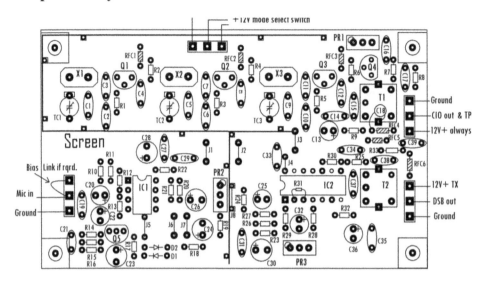

Screen measurements.

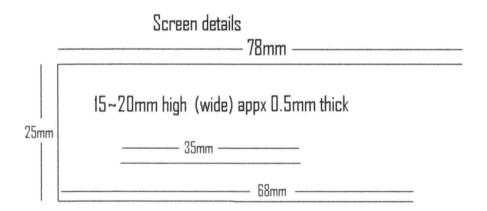

PART 6 MISCELLANEOUS

Component list:

(For 9 Mhz SSB filter)
X1 - 9.0000 Mhz (CW)
X2 - 8.9985 Mhz (LSB)
X3 - 9.0015 Mhz (USB)

Q1-2-3-4-5 - J310 or J104 (n type fet)
Q5   J177 or J176 (p type fet)

D1-2   BA282/BAT85/1N5711
(schottky are better)

TC1-2-3 30pf Cer. 5mm trimmer
T1-2-3 Toko KANK3333R (or 45u0L)
RFC1-2-3-4-5-6-7 150uH mini axial.

10 turn trimmers in-line pins.
PR1 - 220R   (rf level)
PR2 - 5K   (af level)
PR3 - 50K   (car. bal.)

All caps 5mm Disc cer. (Jabdog)
C1-5-9   47p npo
C2-3-6-7-10-11   100p npo
C15-37-42   82p npo

C4-8-12-14-16-17-18
C27-29-31-33-34-35
C38-39-41-43-44   100N

C21   1N
C19   10N

Electrolytic capacitors 3mm pin spacing.
C13-28-30-36-40   100 uF

Electrolytic capacitors 2.5mm pin spacing.
C20-22-24-25-32   10 uF
C23   1 uF
All electrolytics 16V minimum.
NOTE: Capacitor pin spacing allows them to sit flush on the board.

Resistors mini 0.125W rating.
R9-33   47R
R30   56R
R2-4-6-11-19-26-27   1k

R31-36	100R
R8-22-35	150R
R24	820R
R25	1k2
R10	2k2
R18	4k7
R28-29-32	10K

R7-14-15-16
R17-20-21   100K

R12   1Meg
R22
R34 (filter matching) 560R

38/40swg enamelled copper wire
Solder pins.
Material for screens (if required).
— — — — —-

Articles that inspired this project:
Jandek ssb Generator Sprat 76 Autumn 93 (pages 24~26)

Mic vogad: Radio kits and Parts:

For full vogad article see
http://www.radio-kits.co.uk/radio-related/agc_amplifier/index.htm

Thanks to Steve G6ALU at
 http://www.radio-kits.co.uk/
For allowing use the circuit.

Rewinding Toko coils by G4HUP
PDF file download
g4hup.com/DFS/Rewinding%20Toko%2010k%20Series%20Coils.pdf

Component sources:
Jabdog for NPO caps and Toko coils
Ronlin Electronics for 'Press and Peel' film
Rapid Electronics for 0.125W resistors

PCB Proteus software by Labcenter was used to generate schematics & PCB.

QRP SCRAPBOOK

455 kHz BFO for older receivers
Fabio Bonucci, IK0IXI - KF1B, gqrp-italy@ik0ixi.it
G-QRP Representative for Italy

This is a simple 455 kHz BFO I made for an old Hallicrafters pre-WWII receiver, model S-20R. Its original BFO coil (PTO) was broken, so I made a solid state circuit to restore this old receiver. Now it works fine also in CW and SSB.

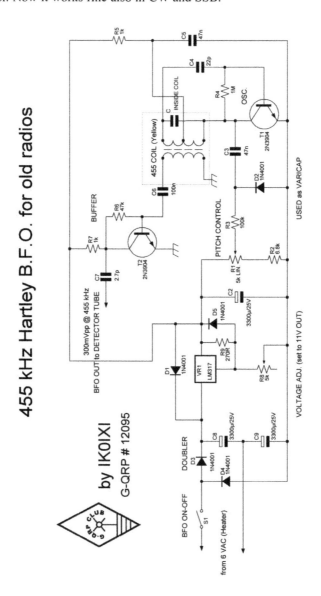

PART 6 MISCELLANEOUS

RF Generation for Superhets
Paul Darlington[1], mØxpd   Pete Juliano[2], n6qw

Introduction
RF generation using Direct Digital Synthesis has been available to commercial radio manufacturers and amateur home-brewers for decades [1]. The DDS devices conveniently are operated under the supervision of a microcontroller, which also provides means to support a user interface. In a recent article [2], mØxpd argued that the introduction of simple "physical computing" platforms such as the Arduino and the appearance of DDS devices on inexpensive modules have transformed these methods from the preserve of specialists to a technology which is i) accessible to all G-QRP members and ii) attractive in terms of price, functionality, performance and simplicity. That argument was presented in the context of direct conversion radios for CW. The present article extends the argument to address applications in a superheterodyne architecture for SSB phone use.

A Practical RF Generation Scheme
A superhet radio involves two signal generators; one to mix the incoming radio signal to an "Intermediate" frequency and a second to mix this to audio baseband. The former generator we shall call the VFO and the latter the BFO. The VFO needs to be capable of frequency change to effect tuning (and, at a coarser scale, band change), whilst the BFO is essentially fixed, changing only to switch between upper and lower sideband operation. In conventional practice, this has seen the BFO implemented as a crystal oscillator, with two crystals providing the LSB and USB frequencies. The system proposed in this article takes the (apparently) extravagant approach of using Direct Digital Synthesis to generate both VFO and BFO signals. The very low price of current DDS modules (e.g. those based on the AD9850 device) makes what once would have been profligacy a rational choice – the second DDS module may easily cost less than a pair of sideband crystals and the resulting system will be very much more flexible.

Consider the system of Figure 1, in which an ordinary superhet receiver is provided with VFO and BFO signals from two DDS modules.

Signals from the antenna are selectively amplified before being passed to a mixer, where they are modulated by the VFO to produce a copy of the desired signal at intermediate frequency. The signal can be processed at this intermediate frequency – typically by a narrow filter, wide enough to pass only one sideband of narrow-band speech. Finally, the processed signal is applied to a second mixer, where the BFO modulates it down to audio frequency. Note that the transmitter uses the same architecture but with the signal flow reversed - literally, in the case of the BITX described below.

---

[1] 8 Uplands Road, Flixton, Manchester, UK   paul@appledynamics.com
[2] 1015 Oakmound Ave., Newbury Park, CA 91320, USA   radioguy90@hotmail.com

QRP SCRAPBOOK

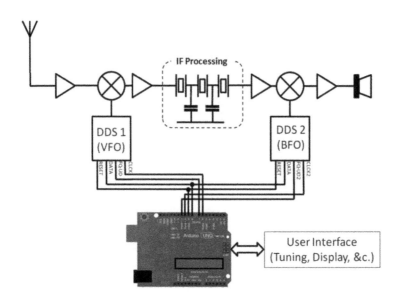

Figure 1 RF Generator Scheme applied to Conventional Superhet Rx

The two DDS devices required to perform this task can be controlled by a simple microcontroller – in the case described here the "Arduino" platform is used (although any other microcontroller family could be substituted at the builder's discretion, if s/he is able to develop the necessary code). The DDS devices have a serial interface, in which serial data on one "DATA" line is defined at the instants of the transitions on a second "CLOCK" line, but there are two further control lines (a data latch command and a reset). Of these four lines, it is possible to share the DATA and RESET between the VFO and BFO, whilst the other two remain unique to allow each DDS to be individually "addressed" by the microcontroller. Thus, a total of six lines from the microcontroller are used to control the two DDS modules, as shown in Figure 1.

Operation of the pair of RF synthesizers is summarised by a single equation, relating VFO and BFO frequencies to what we might call "dial frequency"; that frequency displayed on the tuning indicator.

Dial frequency = BFO frequency + VFO frequency   (1)

BFO frequency is usefully interpreted as a constant, determined by a combination of band and mode. [In practice, for phone operation, it is only determined by band as LSB is used below 10 MHz by convention whilst USB is used above this frequency]. This means that the VFO expresses the variable component of the dial frequency, as is implied in the name <u>Variable</u> Frequency Oscillator. In use, the BFO is set only on band- or mode-change, whereas the VFO is adjusted every time a tuning change is made. It is seen that digital control traffic to the VFO has higher density than that to the BFO.

PART 6 MISCELLANEOUS

When the IF is at lower frequency than the carrier of the signal which the radio is intended to receive, equation 1 is interpreted without confusion. When, however, the IF is above this "target" frequency, equation 1 implies a negative VFO frequency. Although this has meaning as a mathematical abstraction, the DDS module cannot accept a negative frequency input – so the code replaces equation 1 with a pair:

$Dial\ frequency = BFO\ frequency + VFO\ frequency\ |_{Dial\ frequency\ >\ BFO\ frequency}$ (2a)
$Dial\ frequency = BFO\ frequency - VFO\ frequency\ |_{Dial\ frequency\ <\ BFO\ frequency}$ (2b)

The code compares the Dial and BFO frequencies (amounting to considering which band is in use) and selects one of equations 2a and 2b, with the result that the VFO always is instructed to generate positive frequencies!

Advantages
The benefits accompanying use of DDS in the VFO have long been recognised. Most importantly, the radio will offer a level of stability which is hard to achieve with analog VFOs. It will deliver this stability – at high resolution – from below VLF to above HF, which is very hard to achieve with a single analog VFO. There are, however, interesting collateral benefits of DDS available to the home-brewer, especially when using SSB. It will be found that many amateur stations transmit at integer multiples of 1 kHz – due to the digital RF generators in their commercial rigs. Tuning to these stations is suddenly quick and easy when the home-brewer also enjoys digital frequency synthesis and control in their VFO, though the ability to tune continuously between the kiloHertz points remains. The authors note (and, to some extent, share) the reaction of some readers in seeing this "channelisation" as something to lament, rather than as a benefit!

The benefits accompanying use of DDS in the BFO are less familiar; this article's digital, programmable BFO is itself something of a novelty. The ability to tune the BFO will be found of great benefit in the alignment phase of the development of a new scratch-built rig (which usually will feature a homebrewed crystal IF filter of uncertain passband). It allows the BFO frequency to be precisely optimised for the filter and leaves the builder totally free to set IF frequency to any value suggested by available crystals etc.. It even is simple to arrange BFO support for multiple IF frequencies to suit different bands / modes. The benefits available to the home-brew "scratchbuilder" are matched by those which can be accessed by the builder who retro-fits the RF generator scheme here described into an existing rig.

Implementation
Code has been produced to allow readers to experiment with the "Double DDS" RF generation scheme described in these notes. The code, which takes the form of Arduino "sketches", implements both a basic control example and a more elaborated example with conventional "VFO" features (multi-band, RIT, etc).

The overall RF Generator hardware can be assembled using DDS shields, but it is almost as easy to work directly with the DDS modules themselves. The code has been tested on

QRP SCRAPBOOK

various Arduino boards and more experienced users can easily run the code in a stand-alone AVR microcontroller chip. The code and schematics for various hardware configurations are available by following the links on http://www.gqrp.com/sprat.htm .

The authors' experiments with this RF generation scheme have been reported on their respective websites / blogs [3, 4], but key application examples from the contexts of scratch-building and retro-fitting are presented below.

Application: Using the RF Generator in a Scratch-built Rig

mØxpd has been using the RF generator scheme in an SSB rig following the popular "BITX" design [5], allowing multi-band and multi-mode operation. The DDS module produces approximately 1V pk-pk output from 100 Ohms source impedance. This is inadequate to drive the two diode balanced modulator used as detector (Rx) / modulator (Tx) in the original BITX, although the DDS output has been found sufficient to drive the diode ring mixer, which has a driver stage.

mØxpd uses a single buffer / driver to couple the output of the DDS BFO source to the detector/modulator, as shown in Figure 2.

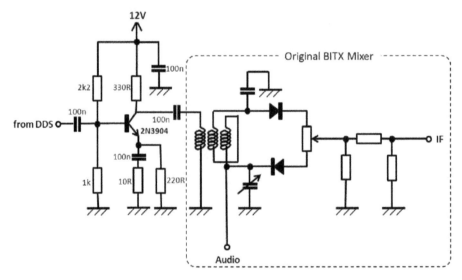

Figure 2   mØxpd's interface between DDS and the BITX detector / modulator

Application: Retrofitting the RF Generator to Existing Rigs

At n6qw there are many homebrew radios and some suffer the general weakness of the lack of an accurate, stable signal source for the LO. With mØxpd's design that is no longer a concern, as several of these radios have been tested with the Arduino/Dual DDS combo and it is like coming out of the dark ages. The first retrofit "proof of concept" application was for the single DDS in a 20m QRP SSB transceiver shown below in Figure 3.

PART 6 MISCELLANEOUS

Figure 3 20m SSB Transceiver with Single DDS VFO

The IF is at 9.0 MHz using the GQRP Crystal Filter and two bilateral amps from G4GXO [6]. The Tx and Rx mixer stage is a TUF-1. (Some cosmetics need to be added to front panel as the 20X4 LCD is larger than the original cut-out.) For good measure I followed the RF generator with a LPF that has a 30 MHz Fc. I do note that the drive is a wee bit shy to the TUF-1 which is a 7 dBm device where you need 1.414 Volts Peak to Peak. As mØxpd noted, the output of the DDS is about 1 Volt pk-pk. A suitable "booster amp" is described below.

This same radio was been tested with the dual DDS RF Generator which is being controlled by an Arduino "Nano". Truly it is a compact package and the underside wiring is done using wire wrap technique, as seen in Figures 4. What a treat to swing down to 40m and automatically be on LSB. Previously, with only the USB crystal available for this radio, going to 40m would be problematic.

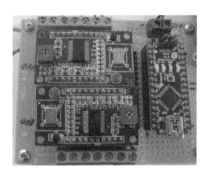 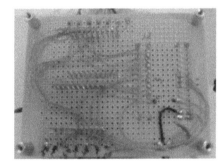

Figures 4 RF Generator implemented using wire wrapping and the Arduino "Nano"

Figure 5 is gain adjustable amp which can be used with a TUF-1 or AE1-L (the latter is a 4dBm device (1.0 V pk-pk)). The BFO requires a similar amplifier.

Another homebrew radio of mine which I dubbed JABOM (Just A Bunch Of Modules) [7] originally used a 4 pole 4.9152 MHz homebrew filter which was replaced by the GQRP 9.0 MHz filter. The JABOM was tested using the dual DDS unit and is a future candidate for a permanent dual system installation for operation on 40 and 20m. Again, the lack of a LSB crystal previously had limited this radio to USB. The total cost, including the Arduino Nano, two DDS modules and 20X4 LCD was under $35, making the price point highly cost effective.

QRP SCRAPBOOK

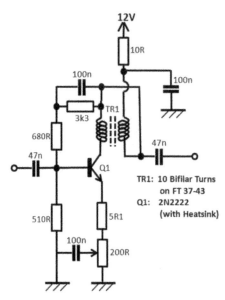

Figure 5 Gain-adjustable amplifier
(after a circuit on the ei9gq website)

One dual unit is slated for installation in a radio that is yet to be built but will use a 3.180 MHz filter from of an early Yaesu FT-101 for which I do not have the matching crystals. Frequently the FT-101 crystal oscillator board is listed on one of the auction sites at a price exceeding the cost of a dual DDS system. Based on the capability of the Dual DDS, I recently bought a FT-101SSB filter for $7.50 USD and now I am ready to proceed with that project.

The dual unit adds a whole new dimension of what is possible with many commercial and homebrew filters. A very exciting opportunity for the dual system is the installation in the boat anchor commercial radios that aside from drifting and the lack of a digital display are otherwise really great sounding radios (valves you know). The dual system also solves another problem frequently encountered with the older radios; their USB and LSB crystals have drifted with age and are no longer useable. The bonus of the dual system is not only a RF generator but the generation of the BFO/CIO frequencies to match the filter.

Using the RF generator in tube type radios must be done with care. A booster amp is definitely needed as is some isolation between the DDS and the radio. I also built an isolated power supply which is fed from the filament string, shown in Figure 6. Avoid powering the Arduino from the plug-in type switching regulators ("wall warts") as the hash generated is noticeable.

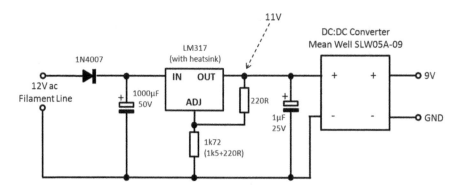

Figure 6 Isolated PSU for Powering the Arduino from a filament string

PART 6 MISCELLANEOUS

The other isolation is the LO RF signal is fed via a ferrite transformer of about 12 turns on the winding to the radio and 4 turns on the DDS side. A FT-50-43 core is used for the windings to better match the radio/DDS. On the radio side one end of the 12 turn winding is grounded and the other end injects the signal through a 10 nF 500 VDC cap. The side of the transformer connected to the DDS is not grounded but acts much like link coupling.

One valve radio used as a test bed was a 1960's Heathkit 40m Single Band SSB transceiver which uses a 2.305 MHz four pole crystal filter in the IF. The Arduino/Dual DDS essentially replaces two valves.

The Heathkit uses a low frequency VFO just above the broadcast band and mixes that signal with an 11.190 MHz fixed crystal oscillator. The resultant difference frequency is now in the 9.5 to 9.8 MHz beating against an incoming signal frequency of 7.0 to 7.3 MHz. Thus the sidebands are reversed but I designated the offsets to accommodate this arrangement. [The IF is 2.305 MHz, the BFO is on 2.303300 MHz and the 'VFO' tunes in the 9 MHz range.] The point of insertion of the VFO was done essentially where the plate of the mixer valve would be connected. The 10nF 500 VDC cap is needed to prevent shorting the LV supply to ground.

The n6qw homebrew radios/commercial radio modifications use a variety of filter frequencies. A few seconds at the computer and these various filters and their offsets can be easily programmed into the Dual DDS code and that is a capability previously unavailable. But that also addresses an issue of how to manage the interfaces and that is the reason for my using the small terminal blocks shown in Figure 2 so that insertion into any project is not a monumental task. Figure 7 shows the Dual DDS Board "al fresco" on the work bench.

Closing remarks
Generation of stable, flexible RF signals for a superhet system can be achieved using DDS modules under the control of computing platforms such as the Arduino. The microcontroller retains the capacity to perform other tasks, such as switching band-specific filters and managing other modes (both of which are planned developments for the code).

Recent "open-source" radio platforms (such as the TenTec "Rebel") attest to the flexibility of and contemporary interest in these methods. We have sought to demonstrate how simple digital methods bring formerly unavailable flexibility to modify our radios literally on the fly. This flexibility is applicable to older "boat anchors" – every bit as much as to modern platforms. The cost of the DDS modules, the accessibility of the microcontroller systems and the performance and flexibility of the resulting system make this technique an attractive choice for those building or retro-fitting radio systems, extending the "Occam's Microcontroller" thesis to phone applications.

QRP SCRAPBOOK

Figure 7   HW-22 & the Dual DDS RF Generator

References

[1] "Pic 'n' Mix Digital Injection System" P. Rhodes, g3xjp, RadCom, January 1999 *et seq*

[2] "Occam's Microcontroller", P. Darlington, mØxpd, SPRAT 156, p13, Autumn 2013

[3] see mØxpd's blog "Shack Nasties" at m0xpd.blogspot.co.uk

[4] see n6qw's website at www.jessystems.com

[5] "BITX - An easy to build 6 watts SSB transceiver for 14MHz", A. Farhan, vu2ese, downloaded from:
http://kambing.ui.ac.id/onnopurbo/orari-diklat/teknik/homebrew/bitx20/bitx.htm

[6] "A 19dB Bi-Lateral Amplifier with 37dB Gain Control Range", R. Taylor, g4gxo, SPRAT 128, p 6, Autumn 2006

[7] "JABOM – A 17m SSB QRP Transceiver", P. Juliano, n6qw, QRP Quarterly, Fall 2011, p 40

PART 6 MISCELLANEOUS

Through a Double Glazed Window
Peter Head G4FYY (g4radioham@gmail.com)

If you don't fancy drilling holes in walls or window frames to get your antenna feeder indoors, then here is a simple solution.

<Fig.1.
1. Shut the window onto a length of bare solid wire to make a template, (fig.1).

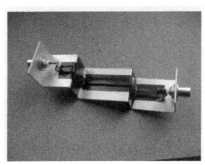

Fig.2.>

2. Cut a strip of aluminium approximately 35mm wide, bend it in a vice to match the shape of the wire and fit BNC connectors at each end. A ribbon cable comprising 4 wires is then taped to the bracket to form an approximation to a microstrip line. Connect the wires in parallel and to each connector, (fig. 2). The characteristic impedance should be close to 50 Ohms but you can easily check this by terminating one end of the bracket with a 50 Ohm load and connecting the other end via a length of 50 Ohm coax to an antenna analyser (or low power TX and VSWR meter). I started with a ribbon comprising 8 wires and peeled off one wire at a time to obtain a VSWR of 1:1 at 30MHz. As shown, the bracket gives a flat VSWR of 1.1:1 to beyond 70MHz and could probably be tweaked for use at 2m.

3. Figures 3 and 4 show the bracket installed in a double glazed window.

<Fig.3 (outside)

Fig.4> (inside)

Please note that this bracket has not been tested at, nor is intended for use at QRO power levels. But if your windows are all the same as mine are, this simple bracket will allow /P QRP operation from any room in the house - hi

WIFB MEMORIAL ENTRY
Simple DDS VFO
Terry Mowles VK5TM vk5tm@internode.on.net

Ever since the appearance of those cheap AD9850/51 DDS modules from China, there have been numerous articles and websites on how to use them. Some of them have been so feature laden to the point of rivalling commercial rigs.

One thing that there didn't appear to be, was a no bells and whistles, down to earth, simple vfo, hence the name of this project.

This project is designed to replace those simple 1 or 2 fet/transistor vfo's that many use for their QRP rigs or experimenting in general. It can even be used to replace the VXO or Xtal oscillator in those rock bound rigs. Tuning was conceived to be over a limited range such as 5 MHz to 5.5 MHz rather than from one end of the HF band to the other.

A look at the circuit will reveal that there is very little to it, just a DDS module, 12F629 8-pin PIC, power supply and support components. All on a 2 inch square pcb. Frequency change is accomplished with a rotary encoder and a toggle switch changes between the two step sizes of 100Hz and 1000Hz (different steps sizes can be programmed if desired).

No provision has been made for filters or amp stages on the pcb as, in some cases it is not needed and in others, the builder may have their own favourite circuit to use. A simple amp or filter can easily be made dead bug style on a scrap piece of pcb material. I would advise installing the Simple VFO in its own shielded enclosure to prevent interference or interaction with other parts of your circuit.

Built into the software is the ability to account for any difference in the DDS modules oscillator frequency via a calibration routine. This needs only to be run once and the updated values are stored in the PIC. Also, the PIC will remember the last frequency used and returns to that frequency when turned on again. To guard against any possible interference from the PIC, it goes to sleep ~ 2 seconds after the frequency stops changing and wakes up again as soon as the encoder is moved.

Construction is straight forward and with so few components, should be completed in very short order. I would suggest, however, doing a voltage check before installing the PIC or DDS module.

There is more information on this project on my website, including calibration procedure and how to calculate new frequencies. The PIC source code and an example Hex file for the AD9850 DDS module set up for a range of 5 MHz – 5.5 MHz can also be downloaded at: http://www.vk5tm.com/homebrew/dds_vfo_8/dds_8.php
Information on using the AD9851 module is in the source code file.

PART 6 MISCELLANEOUS

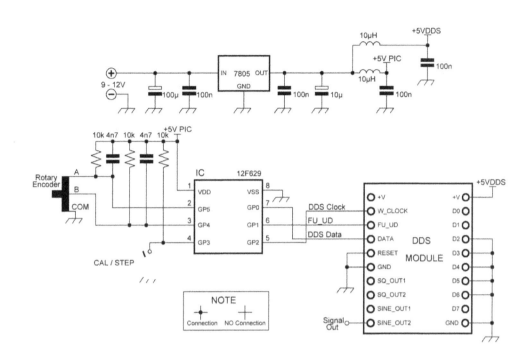

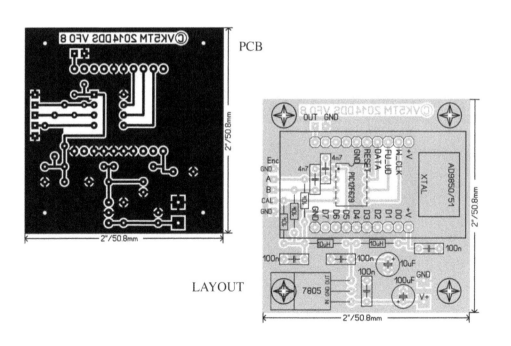

PCB

LAYOUT

HAND GENERATOR for portable QRP
Fabio Bonucci, IKØIXI - KF1B, Email: gqrp-italy@ik0ixi.it
G-QRP Representative for Italy

I made an interesting piece of equipment to recharge my 12V - 7Ah battery.
It's a military hand generator made in '80es by Brownell LTD. for the British military radio set PRC-320. I found it as new on Ebay at a reasonable price.
I think that this generator is an useful accessory, a complementary power source to the solar panels. Thanks to it you can recharge battery also during night, indoor or when the Sun is overcast.
Originally it generates about 24V DC for the PRC-320, so I made a simple solid state regulator to use it on our civilian 12 - 13.8V world. A classic LM-317 is a perfect device for this job. I added a meter to check the charge current (so the state of the battery charge) and a red LED to view when the hand generator reach the optimum revolution speed. I found that 45 R.P.M. is a nice compromise between battery charge and operator fatigue....HI!!

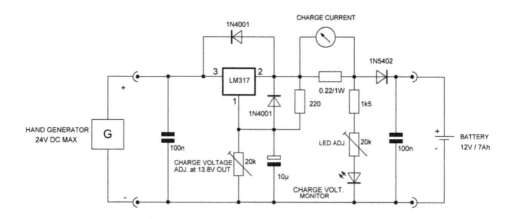

HAND GENERATOR REGULATOR
by Fabio, IK0IXI GQRP#12095

PART 6 MISCELLANEOUS

Two Band Ceramic VXO
Peter Parker, VK3YE, 5/51 Blantyre Ave, Chelsea, Vic 3196
This article was first published in the Sep. 2011 edition of **'Lo Key'** the Journal of the VK QRP Club; our thanks to the club and to VK3YE

Here's a 2-band switched VXO that I've found useful in simple transmitter and receiver circuits.

It comprises two Colpitts oscillators and a buffer stage. All switching is DC so there's no switches to add capacitance and lessen frequency range. In addition capacitor and series inductor (if used) values can be optimized in each oscillator for desired coverage. A two-gang variable capacitor is needed but the sections do not need to be equal.

My own version was initially used in a direct conversion receiver for 80 and 40 metres. This then became a double sideband transceiver. Because I had enough 40 metre rigs, I later removed the 7.2 MHz ceramic resonator and substituted a rare 3.68 MHz ceramic resonator (courtesy David ZL3DWS).

The switched 3.58 MHz and 3.68 MHz oscillators cover nearly all of 80 metres – unusual for a ceramic resonator rig.

Even if you only have resonators for one frequency this circuit could still be useful. For example you could have low and high ranges – the low range for CW use and the high range for DSB. The low range would have a series inductor, larger capacitors for bottom end coverage. A fixed capacitor across the tuning gang would restrict unwanted higher frequency coverage and improve band-spread. In contrast the high range would have smaller parallel capacitors and no series inductor to optimize top end tuning.

Ceramic resonators can drift a little so sometimes it is handy to have crystal control as well. Easy!

Have one oscillator with a ceramic resonator and another with a crystal. You get the best of both worlds and can quickly QSY to the crystal frequency if clear. Eric ZL2BMI uses a similar arrangement and finds that crystal control at both ends allows easy DSB reception on a direct conversion receiver.

All components are easily obtainable. The 3.58 are easily found but the 7.2 MHz ceramic resonators are a little more difficult to source. I used an air spaced tuning capacitor, though even a 60 + 160 pF plastic type should be suitable for a low range/high circuit.

QRP SCRAPBOOK

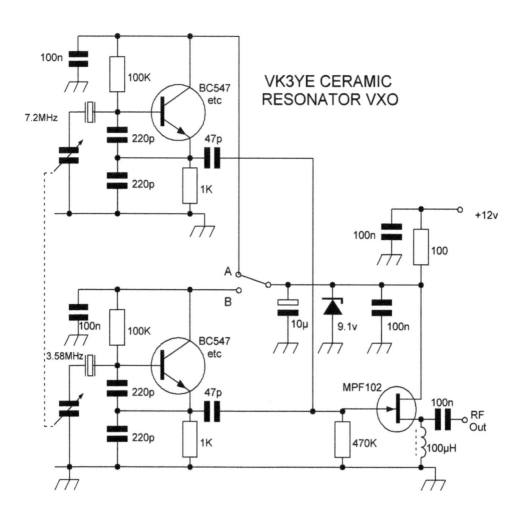

PART 6 MISCELLANEOUS

A 5 Watt Linear Amplifier
By Mike Small G4DVI GQRP Club #302

I needed a QRP linear amplifier for my latest project – a 14MHz SSB transceiver. I decided to base this on the IRF510 power MOSFETs available from the GQRP club. This article outlines what I built and the results that I obtained.

The driver stage in my transceiver produces around 100mW and, using power transistors designed for RF work, it is reasonably easy to obtain 20dB of gain from a single stage. However, getting this amount of gain is more difficult using the IRF510, which was originally intended for power switching at low frequencies. Researching the radio amateur literature on the internet I found a number for existing designs using these devices. Many of these designs were intended as add-ons to existing rigs and so only provided around 10dB of gain and required an input of between 1 and 5 watts. In addition most of the designs were for frequencies up to 7MHz – which is less demanding than 14MHz.

In the end I decided to base my project on the article by Mike Kossor, WA2EBY entitled "A Broadband HF Amplifier Using Low-Cost Power MOSFETs". This appeared in QST March and April 1999 editions. This article describes an amplifier using two IRF510 in push pull with around 50 watts output when run on a 28 Volt supply. I liked the design for a number of reasons: Firstly, I prefer a push pull design because it produces a cleaner output signal, even harmonics are much reduced and this simplifies filtering. Secondly the design provided good performance up to 28MHz with reasonable gain and efficiency. Finally the design is broadband and so would work on other bands.

One of the reasons why it is hard to get these devices to work efficiently at higher frequencies is the high input capacitance (around 180pF according to the datasheet). Many designs try to overcome this with an input step down transformer to use brute force. This is not my idea of QRP! Another approach is to use an input "L" network which uses inductance to tune out the input capacitance at a specific frequency. The WA2EBY design uses inductors (L1) whose value is based on many hours of experimentation to improve the input matching across the whole of the HF spectrum.
The input to the amplifier uses an input balun consisting of 6 turns of 27SWG bifilar wound on a club T37-43 toroid. This is fed from my preamplifier which uses a 2N3866 transistor to provide around 4 Volts peak input in 50 ohms. The balun

matches this to 25 ohms per MOSFET. The matching inductors L1 are made by winding 9 ½ turns of 24SWG wire around a 6mm (1/4 inch) drill. The 27 ohm resistor matches the input at low frequencies and reduces the 'Q' of L1 at 14MHz.

My "downsized" version runs off a 13.8 volt power supply to produce around 5 watts output. This means changing the output matching circuitry from the original design. Using the formula:

$$Rd = SQRT(2*Vp*P)$$
where P is the required power
Vp is the peak voltage at the drain
Rd is the load resistance at the drain

For 5 watts output and 11.2 volts peak drain voltage gives Rd = 12.5 ohms. This impedance needs to be matched to the nominal 50 ohms presented by the antenna and so requires a 4 to 1 step up in impedance. This can be done in various ways including a tuned circuit or a transformer. I have found that the most reliable way is to use two choke baluns – which has the added advantage of being wide band. Each balun isolates the output from the input at RF. The input to two baluns is connected in parallel and the two outputs are connected in series thus doubling the input voltage at the output. (This technique is described in application note AN479 from Motorola). Power is supplied to the drains of the two MOSFETS through a choke comprising 6 turns bi-filar wound on a club T50-43 toroid. The baluns are connected to the drains through 100nF ceramic capacitors. These capacitors must be able to carry the current without a high loss – I used a monolithic type see the parts list for details.

I made these baluns by winding 4 turns of RG158 coax through club T50-43 toroids. At the MOSFET end the two pieces of coax are connected in parallel and the outputs are connected in series. Calculations confirmed that these toroids are large enough to handle up to 10 watts SSB or CW when used this way. With 4 turns the inductive reactance is sufficient for use from 3.5 MHz upwards. However, it is not sufficient for 1.8MHz. As it is not practical to add more turns, if you wanted to use this design for 1.8MHz you would need to use a larger core and more turns.

The output signal is filtered to remove the unwanted harmonics by the filter described in the WA2EBY article. This is a 5 pole Chebyshev design and is suitable for both 14 and 18MHz with an Fc of approximately 20MHz. This is adequate because of the low level of the second harmonic that is a characteristic of the push pull approach.

PART 6 MISCELLANEOUS

Circuit Diagram

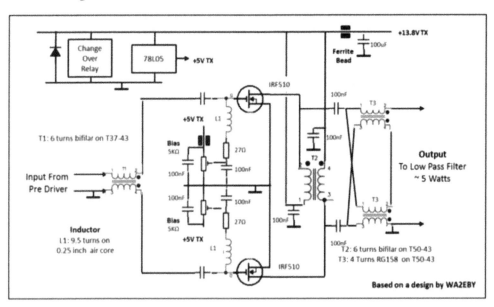

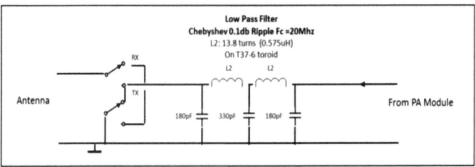

QRP SCRAPBOOK

The two inductors in this are made by winding 13 ½ turns of 27SWG wire on club T37-6 toroids. I used polystyrene capacitors for the filter.

One of the benefits of power MOSFETS over transistors is the simplicity of the biasing circuitry. The temperature compensation required by power transistors is not necessary. All that is required is a stabilized fixed voltage and a potentiometer. I already had a stabilized 6 volt supply that was active on transmit – so I used this as the bias supply.

As an alternative you could use a low power 5 volt regulator IC such as the 78L05 to derive the bias supply from the 13.8 volt supply. This is connected to each gate though an individual 5K preset potentiometer. Adjust each MOSFET to have a standing current of between 10mA and 100mA. I set mine at 40mA per MOSFET.

Another consideration, discussed in great detail in the article by WA2EBY, is the heatsink required. When driven to 5 watts output my amplifier took just under 1amp from a 13.8 volt supply. This is around 14 watts input of which 5 watts goes into the load. So the MOSFETS dissipate 9 watts or 4.5 watts each. To avoid destroying the devices it is essential that they are attached to a heatsink.

I built my amplifier in an aluminium die cast box (111x60x31mm from Maplin). The IRF510 MOSFETs are attached to the box using mica insulators and thermal grease and so the box acts as the heatsink. In practical terms – I ran the amplifier key down at 5 watts output for 10 minutes into a dummy load and the box became warm but not hot.

I built the circuitry on 2 pieces of Vero board – so I don't have a printed circuit to offer.

In conclusion – I am very impressed by this design. It provided sufficient gain for my needs and has proved to be very stable and efficient. I have made several contacts across Europe from Germany to Croatia using this amplifier feeding a 12AVQ vertical.

PART 6 MISCELLANEOUS

Improvements

Further improvements can be achieved by adding a preamp for HF reception and constructing band pass filters for your favourite bands, a good site for calculating these is http://www1.sphere.ne.jp/i-lab/ilab/tool/BPF_C_e.htm

The T820 does contain a preamp but it is noisy (typically 8db) this can be improved especially for reception at the higher VHF frequencies by adding an LNA ideally at the antenna end. If you are not using the dongle for TV then I suggest removing the IR sensor (3 pinned black thing next to aerial socket) as it is a source of noise especially at the higher frequencies. The dongles are best screened by being mounted in a suitably robust metal box. They run hot this is normal.

Station Monitor and Panadapter

These cheap dongles can be a useful second/umpteenth receiver for the shack. They are also useful for monitoring your signal via both the RF and audio spectrum displays.

In addition, if your rig has an IF output, connecting this to the dongle aerial socket means that the dongle can be used as a panadapter by tuning HDSDR to your first IF frequency, then anything you tune on your rigs receiver will appear on the RF FFT display, allowing you to see a panoramic display of the band.

The RTL2832U ADC is only an 8 bit device in the future 16 bit dongles are likely to a appear on the market these will offer an improved performance. If the price is right watch this space.

Have fun with your dongle!

Note: if you want to use a "sound card type Receiver" like the one in SPRAT161 you only need install HDSDR and plug the RX in to the sound card. Follow the user guide https://sites.google.com/site/g4zfqradio/installing-and-using-hdsdr. You only need the zadig driver and the DLL if you are using a dongle. A good sound card helps i.e. one with a bandwidth of circa 50KHz or better still 100KHz.

Screen shot of dongle after HF conversion receiving JT65-HF on 40M

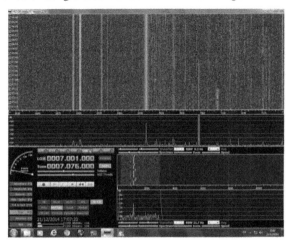

QRP SCRAPBOOK

A real tuning knob for SDR radios.
Tony G4WIF

In the previous Spring Sprat we featured a simple short wave conversion for a "TV Dongle" written by Ken G4IIB. The project turned out to be a lot of fun to play with - but there is something not quite right about tuning around by clicking on your screen. Users of SDR software like SDR-Sharp will have discovered that you can tune by using the mouse wheel - but even that seems like hard work.

I began to wonder if an alternative rotary encoder to the mouse wheel would work better – and it did. Shown below is a typical rotary encoder. Also shown are the innards of an inexpensive mouse showing the mouse wheel encoder.

As you can see, both have three wires, though some encoders have an additional pair of contacts for the shaft "push switch".

I simply connected the new rotary encoder across the contacts of the mouse wheel encoder and plugged the mouse into a spare USB socket on the PC. You can have two computer mice active at the same time on Windows PC's.

You should find that if you move your existing mouse pointer and select the tuning window of your SDR software it will allow the new tuning knob to adjust your frequency. It is also not quite as laborious as tuning using the mouse wheel.

Setting my SDR software to 1kHz steps I found that one rotation of the encoder produced 20 kHz shift, indicating a 20 step encoder but they will vary depending which one you buy. A slight issue is that they aren't all wired the same so some experimenting may be required.

Connect the middle pin of the new rotary encoder to the middle pin of the mouse wheel encoder, then connect the outside two and test. If tuning turns out to be the wrong way around, reverse the outside two connections.

If that doesn't work then the ground (or common) connection must be the first or third pin. Rearrange the wiring and repeat the above test.
Box it all up and you have a simple, "real" tuning knob.

Simple 4 Watt Add-on P.A. for 7MHz Micro Transmitters
Michael Bliss G4AQS.

Needing something to encourage me to go out on my bike I struck upon the idea of some /P operating with a simple transceiver that would fit, along with the other necessary gear, into my rucksack. Looking for something suitable I happened upon a Forty-9er kit from Hong Kong at £8.99 including post! It arrived quickly and was soon built. I have to say the print and parts were much better quality than I expected, however the suggested 3W output was rather optimistic, only giving 700mW. I decided I would like rather more than that but without too much complication.

Having built the rather impressive MKARS80 a couple of years ago and influenced by its P.A. I came up with this very simplified version. It's as basic as I thought I could get away with! It was built on an odd scrap of double sided print, the lands being carved out with a chisel. (Yes, I told you it was basic!) The few components were added and a simple Pi output filter. Feeding it from the Forty-9er and switching on resulted in …nothing…..but turning up the bias to about 0.6V gave over 3W, it really did surprise me, honest. (Make sure bias is 0V before feeding in any RF). The output looked and sounded OK and the IRF510 remained cool even without a heat sink. A better filter was added before air testing and with everything strewn all over the bench, had a first contact with Jos ON6WJ, who was lurking with his Pixie on 7.023.

Like constructing with valves, components seem to be non-critical. The exception being the toroid material in the drain. The number of turns can be plus or minus a couple, the size and mix of the toroid makes little difference but it must be dust iron. Ferrite in any form doesn't give so much output. I suspect there must be a rather broad resonance involved.
To give some idea of its easy going nature: I put 10 turns through an old TV dust-iron IF tuning slug, the ones with the hexagonal hole though their middle and the output was only marginally down! The winding does get slightly warm so I think the bigger the better in that department. It gives a reasonable output when fed 500mW and better than 4W on a freshly charged battery and 700mW in.
Bear in mind that it is suitable only for CW; it's not a linear and there is a drastic drop in output if the key is depressed for a whole minute, although nothing gets hot. It's unlikely that you would do that in practice and it does recover very quickly.
Been too busy in the workshop to get the bike out yet……but….. look out for G4AQS/P sooner or later.

(NOTE)
I have tried to make it reproducible as possible, having varied the resistor values a little. I've tried T37-2; T37-6. T50-2 cores with varying numbers of turns with little effect on output and 3 different IRF510s of dubious pedigree. The bias can be increased without ill effect.

QRP SCRAPBOOK

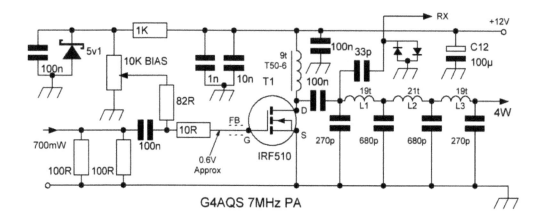

G4AQS 7MHz PA

AUTUMN 2015 SPRAT 164

Follow up on the simple add-on 4W 7 Mhz P.A. (SPRAT 164) Michael Bliss G4AQS.

Unfortunately, when the circuit diagram was transcribed (tidied-up) for publication, the type of toroid cores used in the low-pass filter was missed off. L1, L2 & L3 were wound on T37-6 (yellow) cores using 0.4 mm wire. This came to light when Ian, G4GIR, was constructing one. He tells me that it is now working well, although he did have to use a higher bias voltage; his doesn't fade in power even after 2 minutes key down.

Add-on 4W PA on right
and Chinese Forty9er on left.

PART 6 MISCELLANEOUS

A Switchable Crystal Oscillator
Paul Darlington - m0xpd - 8 Uplands Rd, Flixton, Manchester

I returned home from my first foray into FDIM and realised that the bench needed to be cleared as a matter of urgency. Tidying the bench at mØxpd is something of an archaeological project, in which layers of dirt are removed, artefacts are recovered and a journey back through time is undertaken. During the dig, I unearthed a little circuit that could easily have been tossed into the trash can – but I decided instead that it might be worthy of description in these pages, for several reasons.

Firstly, there was a direct lineage between the circuit and the demonstration hardware I took to Dayton. Second, the circuit is an example of a module – of which I am an enthusiastic user and advocate. Thirdly, the circuit was built on stripboard, which attracted significant interest and comment at Dayton. Last – and most important of all – the circuit, which is an electronically switchable crystal oscillator, has some interest both in itself and in application.

Of these points, the link between the oscillator and the hardware I used at Dayton to illustrate my presentation might be of only passing interest to readers of this article if they were not participants at FDIM. Accordingly, this lineage is addressed indirectly throughout these notes. However, the remaining points are of general interest are considered explicitly in the following sections.

Modularity
I like modular construction. I find that a modular approach helps with design, building and troubleshooting. Particularly, I believe that a modular approach is helpful in an experimental context – so I like to make "building blocks", from which I can assemble larger systems in the pursuit of my RF games (I won't dignify my activities by calling them experiments). Building Blocks remind me of children's play things. They are simple and inexpensive. They can be built up into large structures and then broken down when the child is bored or the structure collapses. The blocks are ready to be pressed into service to build the next project.

I make lots of building blocks – most of which operate on a 12 V supply. They have pins to plug into solderless breadboards, which I use to distribute the power and any control signals and, when there's no other choice, the RF signals themselves. There are RF buffers, RF amps, bi-directional RF amps, small PAs, filters, mixers, AF amps and there are lots of connectors arranged as plug-in modules. You can see them on my blog [1].

The circuit which is subject of these notes was built as a module.

Stripboard or Veroboard ™
My preference to build modules which plug into solderless breadboards via a series of pins generates a requirement for pins on a 0.1 inch pitch. This makes the construction of modules on stripboard a particular convenience, as it is available with both the required

QRP SCRAPBOOK

matrix of holes on the correct pitch and some copper strips from these holes from which to begin the construction of an electronic circuit.

In the UK, such stripboard is produced and sold under the name Veroboard, by Vero Technologies Limited. This prototyping technology was developed in the early 1960s and has been a mainstay of the UK electronics scene – particularly in the "hobby" sector – for as long as I can remember. Judging by the responses and comments the hardware I showed at Dayton received, it is not familiar in the United States.

If you listen to the experts, you will hear that Veroboard and the like is not ideally suited to RF design, given that it is little more than a series of uncommitted capacitors. However, I am much more interested in the practicalities and pragmatism of what works than in what the pessimists and naysayers tell me won't work – so I get on and use what is offered until I find something going wrong and deal with any ugly consequences when they occur. They haven't occurred yet. Of course, that is due in no small part to the fact that I pay some intelligent attention to what some of the "plates" of these uncommitted capacitors are doing!

The Original Oscillator

I had need of a simple VXO circuit to build as a module to run as the local oscillator in a superhet receiver I was developing. A candidate circuit was fresh in my mind - the front end of George, g3rjv's "Sudden" transmitter – and I simply stole that [1]. Here is the schematic and a photo of the first implementation of that simple oscillator as a module.

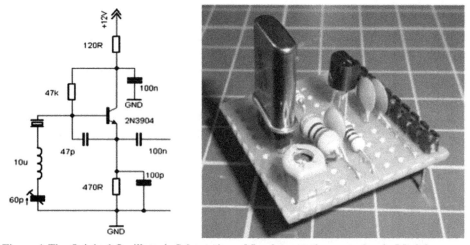

Figure 1 The Original Oscillator's Schematic and Implementation as a plug-in Module

The trimmer and the inductor in series with the crystal allow it to be pulled quite a way from its original resonant frequency, to impart the "VXO" function.

In order to assist in the layout of a circuit on stripboard, a number of software packages are available, similar to PCB design packages. One example *(usual disclaimer – I have*

NO experience of this or any other stripboard layout software) is VeeCAD [2]. These design tools may suit some people's tastes – but not mine. I am interested only in planning how I am going to make best use of a piece of board and documenting – so I have made some graphical representations of common components and an area of the stripboard – to a common "scale" – which I move around in an ordinary drawing package: PowerPoint. The results serve me well. Here is the layout I have used for the oscillator module in the photo above, seen from above...

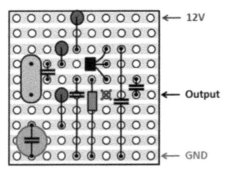

Figure 2   Suggested Stripboard Layout for the Original Oscillator

The hole overlaid with an "x" indicates a break in the copper strip below. Such breaks are made with a special "spot face cutter" tool – the tip of an eight inch twist drill works nearly as well.

The simple oscillator above served well as local oscillator for a superhet [3], but when I wanted to switch the local oscillator frequency in my Parallel IF Scheme [4], it became redundant and I switched to LO generation by digital means.

Before I abandoned the crystal oscillator, I wondered if there might be a way to save this simple technology...

The Switchable Oscillator
We are used to switching a crystal oscillator to introduce a frequency offset between transmit and receive [5]. I was interested to generate a much wider offset than could be achieved by pulling a single crystal - I wanted to switch between two separate crystals. First I tried doing this with FET switches, but enjoyed little success. Then I replaced the FETs with simple BJT devices, and the system of Figure 3 rewarded my efforts.

The base of the same amplifier stage is coupled to two crystals. The other side of these crystals is held at high impedance until one of the transistors Q3 or Q4 is switched on, at which point a low impedance path to ground is established, allowing one of the crystals to resonate.

QRP SCRAPBOOK

The control input to the system is a digital input. When it is at 0V, Q4 is conducting and the oscillator will produce output at the frequency dictated by the crystal connected to Q4's collector. When the control input is taken high (to a 3V3 or 5V logic level), Q3 will conduct and the oscillator will produce output at the frequency dictated by the crystal connected to Q3's collector. By this means, a single digital input line, derived from either a microcontroller or a mechanical switch, can switch the oscillator between two distinct frequencies.

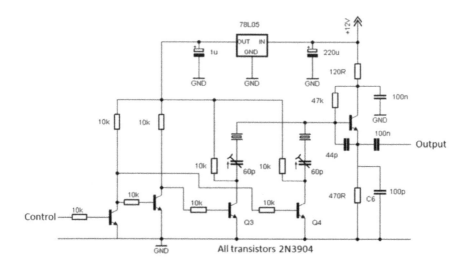

Figure 3   **Schematic of the Switchable Oscillator**

My implementation of this switchable oscillator as a plug-in module is shown in Figure 4.

Figure 4   **The Switchable Oscillator Implemented as a plug-in Module**

PART 6 MISCELLANEOUS

Both the fixed and the switchable oscillator modules used sockets to allow the crystals to plug in, as you see from the photos in Figure 2 and 4.

The particular layout I used for the switchable oscillator is shown in Figure 5.

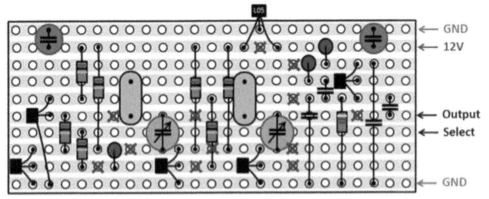

Figure 5   **Suggested Stripboard Layout for the Switchable Oscillator**

I did not include the series inductor in the switchable oscillator – there wasn't space either in the schematic or in the build on the little piece of stripboard. The inductor is valuable for, as has already been mentioned, it increases the range over which the resonant frequency can be adjusted. If you want to add an inductor, place one between each crystal and its trimmer capacitor, NOT between the crystal and the base of the final stage transistor.

Application
The switchable oscillator was developed with the idea that it could serve in my parallel IF superhet scheme, avoiding the apparent extravagance of a second DDS generator. This would have been a worthy application, but for two limitations, which diluted the original motivation – even for a cheapskate like me...

Firstly, although the Parallel IF scheme uses two LO frequencies to change between narrow and wide receiving bandwidth in LSB, which the switchable crystal oscillator described here could deliver, it uses four LO frequencies to handle narrow and wide receiving bandwidths in LSB and USB. The switchable crystal oscillator could be extended to have four crystals, supporting four frequencies of oscillation. But by then it really would be more practical to use a DDS.

Second, whilst the use of a second DDS module seemed extravagant back in the day when we were all using the AD9850 for RF Generation [6], multiple channels of independent RF generation suddenly became easy with the advent of the Si5351 device, particularly on platforms like my Si53531 Shield for the Arduino, available through Kanga UK.

Forgive me – I'm biased.

QRP SCRAPBOOK

In the face of such competition, the switchable oscillator described here is of little interest for the Parallel IF system – but perhaps it still has use for sideband switching in an ordinary superhet. Or perhaps it is only of archaeological interest!

See you at FDIM 2016.

References

[1] http://m0xpd.blogspot.co.uk/2014/08/ocm.html
[2] http://veecad.com/index.html
[3] http://m0xpd.blogspot.co.uk/2014/08/breadboard-bitx.html
[4] Darlington, P. (2015). A parallel filter architecture for flexible IF processing. RadCom. 91 (1), p78-83.
[5] Hayward, W, Campbell, R & Larkin, B (2003). Experimental Methods in RF Design. USA: ARRL.   p 6.65.
[6] Darlington, P & Juliano, P. (2014). RF Generation for Superhets. SPRAT. 158, p4-11.

Low Cost Adjustable PSU
Mike Bowthorpe, G0CVZ.   mike@czechmorsekeys.co.uk

The item is still listed on ebay

I have just built a low cost adjustable PSU using an eBay item see 261268313276

I attach a photo, all I have done is boxed and fitted sockets, but the input socket can take a laptop plug, so when you need the PSU just borrow it, adjust to suit and away you go ~ cheap, accurate, with power source that has more than 1 purpose :-)

The PSU is 0 to 40V and 0 to 2A out depending on what you feed into it. It says it can provide 3A, but I have not added any additional heat sinking

The Wireless Side Tone
Richard Wilkinson G0VXG   richardwilkinson@aol.com

I was asked to make up a 40m Pixie transceiver kit that a friend bought off the internet for £3.00. The kit worked as well as could be expected but I found it very difficult to send CW without a side tone.

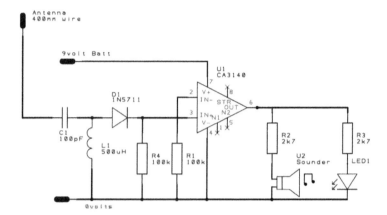

I considered just adding a piezo sounder to the board, but after some thought decided that it would be useful to have one that could be used with various QRP transmitters.
There are lots of circuits on the net (Ref1) for RF sniffers but I found that most suffered from mains hum being induced into the circuit. In the end I had to add a simple filter C1 L1 which is well above 50Hz and its harmonics. The CA3140 is a very useful IC having a high input impedance and operates from a single rail 9volt supply.
The power consumption when idle is about 1mA and goes up to 5mA when the sounder and ultra-bright LED are on, so a PP3 battery is fine. If a visual signal is not required then the LED circuitry can be omitted. The 2k7 resistor works well with my 3volt sounder but can be changed if other sounders are used.
 For testing, the Pixie was powered with 12volts and connected to a 40m dipole and by just placing the coiled 400mm antenna wire next to PA the transistor, the side tone could be heard clearly. For more tests I used the MFJ antenna analyser which only outputs a low level RF signal. The frequency goes from 1.8MHz to over 400MHz and the side tone worked across this range.I am sure this will be a useful aid when at Rallys and junk sales to give some confidence that a Handy or RC device is at least outputting some signal! The piezo sounder that I used does not have the best tone in the world, so maybe a sinewave oscillator could be added if required.

Please note that the op amp is cmos and the usual precautions should be taken.

Ref1 http://www.circuitdiagram.org/long-range-cell-phone-detector.html

QRP SCRAPBOOK

Economy iambic keyer –
Andrew Keir, G4KZO, keirfamily20@gmail.com

I like sending iambic morse but the cost of 'bought' keys is frightening. I wanted something cheap, easy to make and simple to operate. This is my mod of a standard 'sideswiper' pcb key, with 2 double-throw microswitches and offcuts of glass fibre pcb material (preferably double-sided). It doesn't have 'tension' or 'swing' control but do I actually need them?

The 2 x spdt microswitches are wired to the 'normally on' terminals, with the twin long strips of pcb material holding them 'off' (depressed) until sideways movement of the strip allows either microswitch to return to 'on'. I used a base of pcb material about 70 x 85 mm with the two levers 100 mm long and about 10 mm apart; you may wish to vary this.

The supports are best made from double-sided pcb material, and soldered on, but of course could be glued. I had solder available, and it can be reworked, so I used that. It is important that the levers do not 'brush' the base, so I put a sheet of paper between them while lining up and soldering the levers. This guarantees about 0.5 mm space between levers and base, when the paper is removed, after the solder has set.

Once the supports and levers were right, I put the microswitches roughly in place and 'spotted through' one of the mounting holes on each microswitch, drilling a hole in the base to suit the pivot bolts (I used 3 mm but 6 BA would do). A nut holds this 'support' bolt tightly to the base, and the microswitch pivots freely on it. It should not be necessary to use a second nut to clamp the microswitch down, as it does not move in normal use. A second bolt, through the other mounting hole, holds the switch level.

Now solder on 4 pieces of scrap, to hold the side adjusting bolts for the microswitches. Drill (and, I suggest, tap) the outer supports for the bolts; fit the bolts (with nuts outside and inside) and adjust, until they hold the microswitches against the levers, 'clicked' but only just.

Moving either lever inwards 'un-clicks' the microswitch, switching it 'on' from its forced 'off' position. I managed to get an operating movement of just less than 1 mm, which suits my sending style. I hope it suits yours.

You will need –
- 6 x 3 mm dia x 20 mm long bolts, and at least 6 nuts to fit them
- 6 x doublesided pcb material scraps, about 15 mm x 10 mm
- 1 x pcb material scrap, about 30 mm x 10 mm
- 2 x pcb material levers, about 100 mm long and 10 mm wide
- one pcb material base, about 70 mm x 85 mm
- a heavy chunk of something to bolt/glue the base to, to avoid 'wandering key' syndrome
- 2 double-throw micro-switches; the ones I used had a base about 25 mm x 15

G4KZO Key

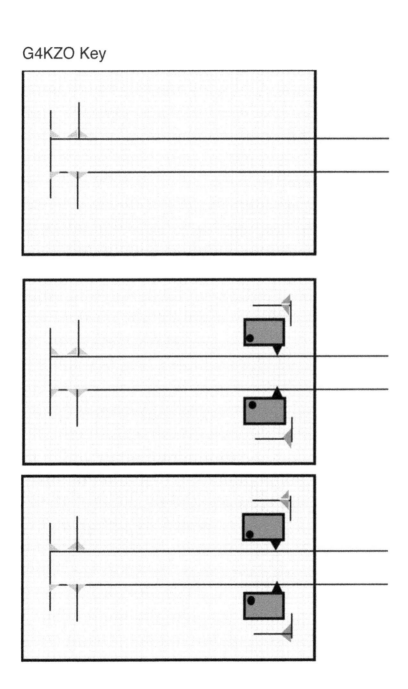

QRP SCRAPBOOK

CQ Keyer
Peter Howard G4UMB 63 West Bradford Rd Waddington Lancs

If like me you make QRP transmitters that are mainly crystal controlled you will have experienced the long wait when calling CQ lots of times to get a contact. This simple Iambic type A or B keyer IC is controlled by software designed by Wolfgang DL4YHF, which can be downloaded free of charge from the internet .

The two push button switches allow you to input numerous commands so you can customise the keyer to your requirements. In a simple setup which is adequate for me I have only used one button on pin 12. Pressing this button for more than 2 seconds will command the software to flash the letter M on the LED. You then can record a short message using the paddles and after that press the button again for 2 seconds to store the message The LED flashes an S. This message will be stored even if the power is disconnected afterwards. Touching either paddle during the playback of a message will stop playback and return the keyer to normal.

I have drawn this circuit which is only a small part of what the keyer really is to show a typical application for a QRP transmitter using it to drive a relay and piezo sounder . A typical message to store would be a CQ CQ call   Go to this website for ALL the info.
http://www.qsl.net/dl4yhf/pic_key.html

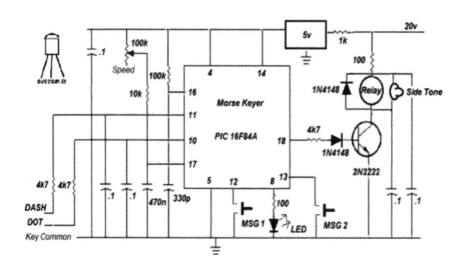

PART 6 MISCELLANEOUS

Using ceramic resonators in SSB filters
Bernie Wright, G4HJW , bernie@earf.co.uk

Ladder filters using low cost quartz resonators are well known in amateur circles, but what about ceramic resonators? Actually, they can be quite useful in lowish frequency SSB filters, despite their lower Q. The reason for this is that the lower impedance makes top coupled parallel resonant operation a suitable configuration and one which I think makes for an easier implementation. Associated network capacitors end up being convenient values, and 'suck it and see' filter development (ie, trimmers initially used in all capacitor positions) make the whole thing quite fascinating to experiment with. In the example shown, three 455 kHz resonators are used. The shunt capacitors allow some movement of resonator frequency, whilst the series capacitors control the coupling between stages and have a major effect on the filter shape. With this topology, it is also easy to arrange for 50 ohm input and output matching impedances. This particular filter was fitted into a small 80m monitoring receiver, the audio from which certainly sounds nice.

Other sideband filters have been built between 300 kHz and 1 MHz and with up to five resonators in the case of the 1 MHz filter. Sometimes, the loaded Q of one of the resonators may be found to be a little high. In these cases, damping resistance can usefully be added across the particular section to flatten the overall in-band response.

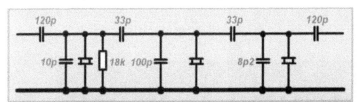

Resonators at 3.58, 3.68 and 14.32 MHz can be used in the same configuration to provide receiver front-end selectivity over about a 20 – 30 kHz bandwidth for a typical pair of resonators.

For more graphs and circuits, see http://www.earf.co.uk/cerfiltrs.htm

Bernie Wright
G4HJW

QRP SCRAPBOOK

Add-on Squelch
Peter Howard, G4UMB, 63 West Bradford Road Clitheroe Lancs BB7 3JD

I have been listening to AM and FM using my homebrew 160m80m AM receiver (See Sprat No.138 page 21) but have been annoyed by the amount of QRN it picks up lately in comparison with a proper commercial rig which has FM with squelch. So I built this simple add-on Squelch circuit to mute the QRN on my receiver. It's a circuit that I had to tailor to my requirements so if you build it and are not bothered about messing with your receiver you will need to experiment where best to fit it. It needs to go after the detector and before the volume control and have at least 2k of resistance between the input and output so that the muting effect does not overly reduce the wanted input signal. To set up you must carefully adjust the preset resistor so the circuit is at the right sensitivity.
Instead of having a Squelch level control like a proper receiver does I found that adjusting the RF gain control on my receiver gave the same effect. FM and AM are the best modes to use it with. but if you use squelch on CW reception it does work and sounds similar to auto break in. The circuit is just a straightforward AF amp using a 741 IC The signal is then rectified into a DC voltage that powers the BS170 transistors. I chose the muting switch as a reed relay to stop any sound breaking through the circuit. I have included a switch in series with the mute switch to turn it off. The circuit has been useful to me in listening to FM on 160m using my simple AM Homebrew receiver by slightly off tuning to the edge of the signal to achieve a form of FM detection.

PART 6 MISCELLANEOUS

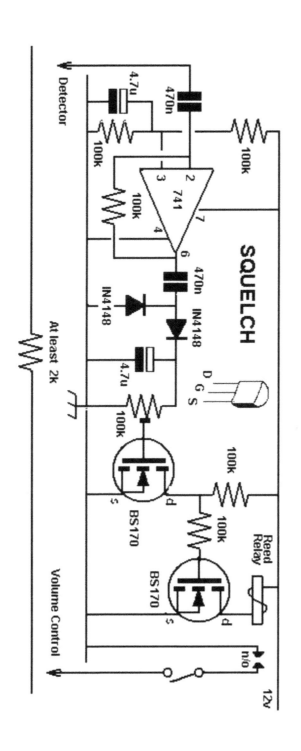

SUMMER 2016 SPRAT 167

QRP SCRAPBOOK

EASY 4-6 CONVERTER
Alan Troy G4KRN   alantroy49@gmail.com

Many modern HF transceivers cover 6m but not the 4m band and here is a receive converter that will give 4m receive on 6m . This should be useful for monitoring any local or semi-local activity in normal conditions but is not suitable for anything more serious as 6m breakthrough is likely when that band opens up. The circuit is based on typical converter circuits from the 1980s that used a dual gate Mosfet but replaces this with two FETs in cascade. (see George's "Carrying on the Practical Way", Practical Wireless, June 2006).

All three FETs are 2n3819. The oscillator uses a 20 MHz crystal obtained from Club sales. Both coils are air core, self supporting, using 22 swg wire and centre tapped. Diameter about 8mm. They can be wound on a biro initially to get the diameter. Coil L1 (4m input) is 5 turns, centre tap and coil L2 (6m output) is 9 turns, centre tap. The coils may need stretching or contracting for best results. The RFC in the oscillator is 10 turns 32 swg on a ferrite bead. Do not forget the 47pf capacitor in the 6m output which blocks 9 Volts getting to the receiver. Note in the circuit diagram the drain to L2 line crosses over the gate to 100k resistor, it is not a junction. Make up the oscillator section first and check this puts out 20 MHz. Once the converter is made up, an oscillator on 7 MHz can be used to put out a tenth harmonic on the 4m band as a test signal.

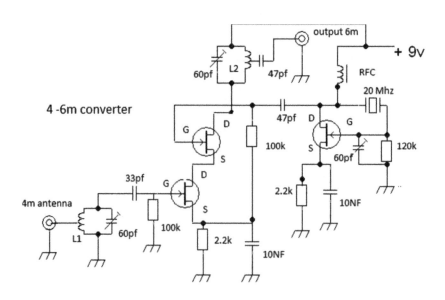

Overvoltage protection using MAX6397
Robin Harris, G4GIY, 303 Northate, COTTINGHAM, HU16 5RL.
robin@sprite.karoo.co.uk

I came across the Maxim MAX6397 device recently and thought it had some useful features. Essentially it monitors a supply voltage and if this exceeds a preset threshold it very quickly disconnects the load. The switching element is an N-channel MOSFET - the voltage drop is negligible and any power level can be switched by choosing a suitable MOSFET.

The threshold is set by using two resistors in a potential divider. The maximum input protection is up to 72V. Switching times are of the order of 100uS so plenty fast enough to keep downstream equipment safe.

This is the circuit:

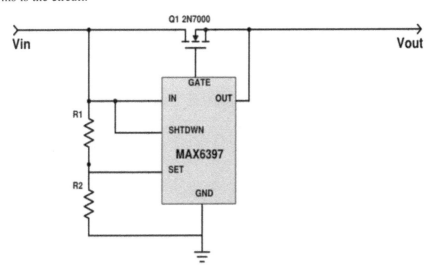

I used a cheap 2N7000 n-channel FET but other choices could provide more power handling.

R1 & R2 are chosen to set the threshold. There is a full description in the datasheet but this approximation is fine:

Rtotal = (R1 + R2) should be about 1 - 2M
R2 = 1.215 x Rtotal / Vov   (Vov is the desired threshold voltage)
R1 = Rtotal - R2

I used a pot to experiment with the threshold voltage and then used fixed resistors in the final circuit.

QRP SCRAPBOOK

Things to be aware of:

- The IC is available from Mouser (UK) at £3.36 (May 2016) for a single device
- The package is SMD 3mm x 3mm so mounting requires a specific PCB pattern. I bought mine from Proto-Advantage - they are not cheap with postage from the USA but it is really the only way of safely mounting the chip. For my first attempt I soldered thin wires directly onto the IC but it was VERY tricky and only suitable for testing and not for a real application.
- Be careful buying a PCB adapter - this IC uses 0.65mm pitch so the part number from Proto-Advantage is IPC0063.
- Soldering using paste and a hot air gun worked fine
- Pin 2 is labelled SHTDWN and must be held high to turn on the IC. Connect it to IN for normal operation. This pin can be used to control power with TTL logic levels.
- The IC includes thermal protection and will shut down power if it overheats

The IC also includes a 100mA 5V regulator and a pin (POK) that monitors this level.

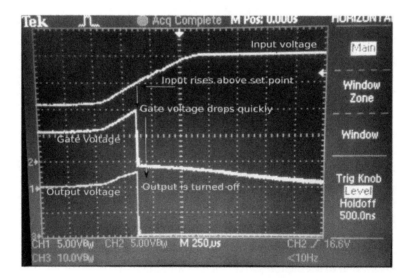

The screenshot above is an oscilloscope trace to show how the device responds to an input voltage over the threshold. At that trigger point the gate of the FET is taken low isolating the downstream equipment from the supply. When the overvoltage condition is removed there is a short startup delay before the supply is reconnected to the equipment.

PART 6 MISCELLANEOUS

A Specific 5V Option – LTC4360

Another alternative for protection of 5V systems is the Linear Technology LTC4360. This comes in an 8-lead SC70 package and will disconnect downstream equipment when the supply exceeds 5.8V. It is rated for up to 80V input.

This device also switches an external n-channel MOSFET - I used a 2N7000 again. The circuit is very simple but remember the ON pin must be grounded to turn on the output. This pin can be used as a switch - take it above 2.5V to turn off the output.

The PWRGD pin is held high by an internal pull up resistor and goes low when the output is connected i.e. 'on'.

Soldering is a little easier with this device as it has 'legs'. I used a PCB adapter, solder paste and a hot air gun and had no difficulties.

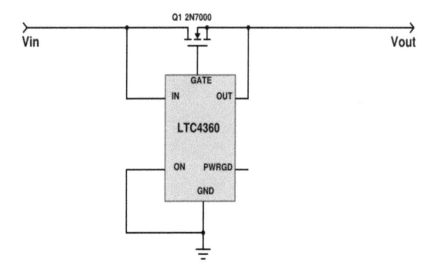

QRP SCRAPBOOK

Cheap and Cheerful Single Lever Paddle
Herb Perkins WA2JRV
WA2JRV@gmail.com

While I must give credit for the overall design of the paddle to Dave AA7EE, I did build one based largely on his design.

As I am quite cheap (read that as frugal) I took apart an old IDE hard drive that was just a door stop, removed the magnets, broke one into four small sections and used marine epoxy to put four small pieces on the respective corners of the mounting board. I also covered them with epoxy so the FT 817 would not get scratched.

The QRP Guys (http://www.qrpguys.com/) offer the Paddle kit for $15.00. The pc card material was scrap, the hardware was from the junk box and the cable to the radio was from an old set of ear buds that were well past their prime.

For a junk box paddle, it has worked well for me.
73's
Herb

Cheap alignment tool for slug-tuned coils
Gereon Ostermann DJ1WY Hauptstr. 35 D-55568 Staudernheim. GERMANY

Ever needed to tune a ferrite slug in a TOKO-coil, for example tweaking a band-pass filter in your latest project?
Well, at least DJ1WY had to do this a few times in many years...not enough yet to justify purchasing a professional non-metallic "slug alignment tool", though. Instead of this I always used the end of a wooden matchstick. With a little whittling the need tip is quickly prepared at the bottom-end of the matchstick, and it is solid enough to last at least two full tuning procedures. After that the match still can be used for its intended purpose. Fast, easy and cheap.

PART 6 MISCELLANEOUS

Observations with Ceramic Resonators In Super VXO Mode
Mark Dunning VK6WV

I have used ceramic resonators as VXO's for some years and have found them a convenient method of obtaining a stable range of variable frequencies, although this is limited to a restricted range based upon standard frequencies unless one resorts to mixing. Recently I have seen reports of the use of super VXO ceramic oscillators (two or more resonators in parallel) with a very wide tuning range resulting. This was attractive to me so I thought I would do some investigating.

When I first tried paralleled ceramic resonators about a year ago I was not able to replicate the reported results. I was very busy at the time. I assumed that I must be doing something stupid and put my circuits aside until I had more time. When I finally did get back to testing I confirmed that rather than parallel ceramic resonators increasing tuning range, the reverse occurs, the tuning range is reduced. This is the reverse of my experience with crystals operating as a super VXO and I have not seen this reported in the literature.

I tried various popular circuit configurations but they all had the same result, although the various oscillators did affect the nominal centre oscillation frequency and range available in tuning.

I decided to record the results and list them below in table 1. For simplicity of testing I chose to use the Colpitts oscillator circuit shown in figure 1. I did try adding various values of chokes in series with the resonator to see if this would improve the situation. It did not. The nominal centre frequency was shifted by the choke but it did not improve tuning range. In addition I found that when I added a choke in series with the resonator, if the choke was large enough, over part of the tuning range the resonator lost control and the

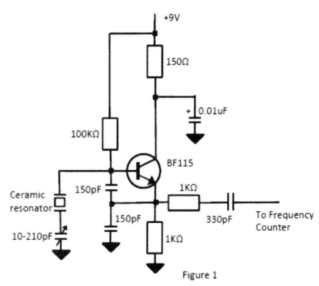

Figure 1

oscillation frequency became dependent upon the inductance of the choke and the sum of all the circuit capacities, in other words a free running VFO rather than a ceramic resonator controlled VXO. The reason for this became apparent when I measured the parallel capacitance across the ceramic resonator. It ranged between 28 and 54pF depending on the resonator. With this range it was not hard to see how the VFO came into action once the ceramic resonator lost control.

Nominal Frequency	Resonator Parallel Cap (pF)	With Minimum Cap (MHz)	With Maximum Cap (MHz)	Difference %	Number of Resonators
4.00MHz	49.6 *	3.98091	3.90117	2.04%	1
		4.00106	3.93909	1.57%	2
4.19MHz	38.2	4.23036	4.11949	2.69%	1
		4.26589	4.16320	2.47%	2
		4.26840	4..18639	1.96%	3
4.91MHz	38	4.95990	4.84639	2.34%	1
		4.99201	4.88579	2.17%	2
6.14MHz	52.7	6.22658	6.02677	3.32%	1
		6.26541	6.09696	2.76%	2
7.16MHz	31.8*	7.11340	6.91339	2.89%	1
7.16/7.20MHz		7.18415	7.00527	2.55%	7.16//7.20
7.20MHz	27.9*	7.147277.	6.95512	2.76%	1
		7.20311	7.02229	2.57%	2
8.00MHz	40.1	8.09332	7.85342	3.05%	1
		8.16341	7.93564	2.87%	2

Tuning C Min	10.5pF	All resonators two lead types apart from those with *
Tuning C Max	208.9pF	Capacitance measurements made with AADE L/C meter IIB

I think that the data can be interpreted as follows;
The resonators are operating in a frequency area where they act inductively in the oscillator. When two or more are in parallel it has a similar effect as might be obtained by placing inductors in parallel. That is the effective inductance reduces so the nominal centre frequency of oscillation increases. This itself can be a useful effect, particularly in the 40m area.

(Continued on page 222)

PART 6 MISCELLANEOUS

SLA CHARGER
Brian Harris G3XGY

I have a couple of sealed lead acid (SLA) batteries that I use in the shack when operating QRP with dc receivers. I know it is important to charge with a constant current with an eye for the maximum time. The circuit I use is based on the positive temperature coefficient of a car side lamp.

The supply was from a laptop charger from a Dell 1150 to give me 19V at 4.62A. The diode is necessary to prevent the possibility of the charger being rendered unserviceable by the battery when the charger is disconnected from the mains supply. The fuse is perhaps unnecessary, but provides an anchoring point for the diode. Barreters are difficult to get hold of and I could've used a thermistor. However, the car lamps provided me with some flexibility and I could, at least, see what was going on. The 3mA meter is ex-19 set, I believe. I adjusted the voltage multipliers and meter shunt values to suit the scales, so I could read 15 and 30 volts, and 3 amps full scale. I used a digital multimeter as my standard and I was happy with a 10% tolerance. It was fun anyway and took me back to my early days of training. The resistance values shown are only examples and the meter shunt is hook-up wire wound on a large resistor.

In use, one 12V/21W lamp gave me 700mA charging rate which suited my 7Ah SLA and two lamps in parallel gave me a 1.2A rate for my 12Ah SLA. These rates would be for C/10, and to allow for efficiency losses, allows for a charge time of 15 hours. Lower powered lamps can be used to charge smaller SLAs.

My thanks go to Mike G3PCW for help in the circuit design.

SLA CHARGER - G3XGY

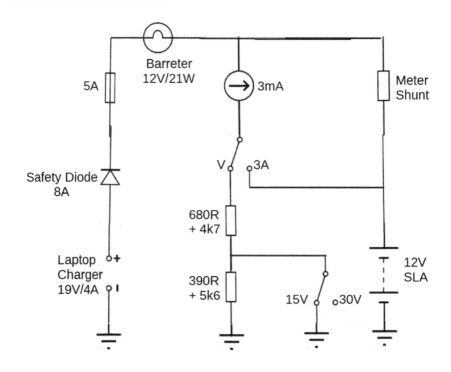

(Continued from page 220)

In my case the amount of frequency change using the variable capacitor that I used is in the order of 1.5-3.3%. This amount of change reduces a bit when the oscillator has more than one resonator in circuit. This is totally different from what I have read on the web.

I think that the tuning range of paralleled resonators is reduced because of the high capacitance across the resonator. I think with multiple resonators in parallel this high parallel capacitance becomes a dominating effect effectively reducing the tuning range of the variable capacitor. By the way the parallel capacitance of the resonators seems to be temperature dependent as well which is not a good thing. This high parallel capacitance may also cause problems with unwanted oscillation modes in some circuit configurations.

Check your key practice
Bernd Kernbaum, DK3WX, Ruppinstr. 13, 15749 Mittenwalde Germany
(dk3wx@darc.de and http://www.dk3wx-qrp.homepage.t-online.de)

If you have problems with single or squeeze paddle or Mode A or B Keyer analyse your key practice. With a Digital-Multichannel-Recorder you may save the keyed text and analyse it. I have not such an expensive device so I build a simple weekend version. One channel of a stereo recorder saves the result of keying (side tone). The other channel saves actions on the Dot and Dash Line. This control channel is amplitude coded and shows the paddle line activity.

Paddle pressed	Amplitude
no	0
Dot	0.3
Dash	0.6
Dot + Dash	1

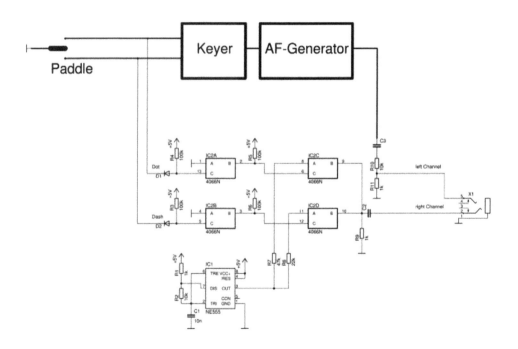

Fig. 1 Circuit – Add a few parts to your keyer and sidetone (dark line)

QRP SCRAPBOOK

An audio generator with the well known 555 IC generates a 6.5 kc square wave. With an analog switch 4066 and a voltage divider R7 – IC2C – R9 you get a small amplitude to the audio line input if you press the Dot-Line-Paddle (0.3). Press the Dash-Line-Paddle and the R8 – IC2D – R9 voltage divider increases the amplitude (0.6).
The analog switches IC2A and IC2B act as inverter with a high resistive input.
All parts are placed on an universal PCB and wired in ugly construction.

I am using the FREEWARE program Audacity for an audio recorder. Figure 2 shows the control channel with the pressed dot line for a short time and subjacent in the other channel the dots. The middle part shows the pressed dash line with the greater amplitude and the dash sequence. In the third part both paddles are squeezed, the amplitude of the control channel is still higher than the other and a dash dot sequence is generated. You see I pressed the dash line first.
For better comparison you can zoom in with the + button.

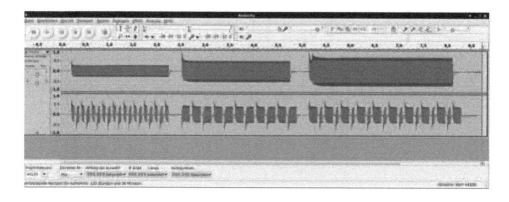

Fig. 2 Audiorecord by Audacity

There is a transient effect when the square waves start but I think thats not a problem. Start with the first edge.

PART 6 MISCELLANEOUS

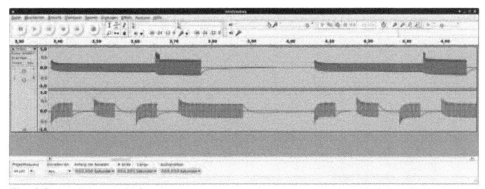

Fig. 3 Record the symbol „v"

I key a few vvvv. First the dot line is hold, during the last dot the dash line comes along. A short moment later only the dash line is active until the middle of the dash. The overlapping Dot/Dash is shorter in the second v.

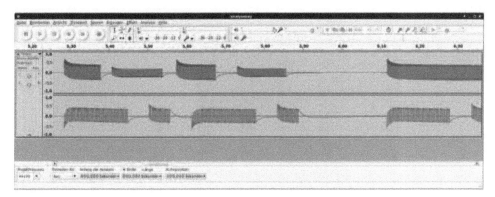

Fig. 4 CQ call

The CQ call whistle-blow - I am not a good squeezer.
In my last project I had problems with the keyer program and I hope this little circuit can help to find a solution.
I have no experience by training a good squeeze mode but I will try and learn. If you be aware a better method please let me know.

72 Bernd DK3WX

QRP SCRAPBOOK

A Novel Morse Key
Peter Howard G4UMB

Here is a very simple method of using a micro switch with a lever soldered to a Phono plug as a QRP Morse key. Ideal for portable operation

Extra Tip G1CXE

Sitting contemplating the fact that a watched kettle never boils, on the way back from a rally, I noticed that the red plastic cap from the small gas cylinder made a nigh on perfect push on cap for keeping dirt out of N-type plugs, and at an effective cost of zilch! An alternative design from the same type of cylinder fits the sockets beautifully.

PART 6 MISCELLANEOUS

A simple, cheap S-meter for portable rigs
Steve Collins, M6SLO

One of my reasons for starting up amateur radio was to try and pick up some electronics knowhow, as a foundation licence I can't use homebrew stuff, but I can still make it, so I have been doing just that. I decided that I wanted a PFR3B, I had always liked the look of that rig, so I bought one of the early 3B's and set to building, I'm in the process of upgrading to the intermediate so I can use it. Along the way I added an external connector to the battery pack so I could monitor the voltage and a little LED log light on the front panel, but what I felt was missing was an S-meter.

Those of you who have seen one of these rigs will know that there isn't the space to put an old CB meter in, so I figured it would have to be electronic/LED based. One option is to use an LM3915N, a ten channel/level device, but you need to get them from Farnell, and I wanted something using easily and cheaply sourced components. Having searched the web I found that Mike Martell, N1HFX, had made a level meter using an LM339N quad comparator[1], and these could be sourced cheaply from Maplin.

I built Mike's circuit up on breadboard and found that it worked well for DC voltages, but not AC voltages, this matters depending on how you intend to derive your "signal", if you use ALC voltage, then it's not an issue, if like me you don't have that option a slight modification is required to measure AC voltages.

The next issue is the level of voltage used as a signal, in many cases this will be too low for the comparator to work, so some sort of pre-amp is required, mine was lashed together using a 741 opamp, yet again easily available. The final polish was on the potential divider that sets the "levels", Mike had just fed this straight from the supply line, however I thought that I'd run mine from a 5volt regulator, as this was a battery powered rig, and the supply voltage would drop over time, I used a 78L05.

A circuit diagram is shown below, with component values, you will probably need to alter the potential divider levels, and possibly the amp gain, dependant on where you derive your signal, my pick off was before the volume pot on the PFR3B, which meant that I lost the amplification of the LM386, hence the need for the preamp.

I feel sure that I have made some fairly big mistakes with this, but, it does work. A slight hum can be heard through the 'phones, but its not intrusive, this was made up onto a small piece of vero-board, the solder side potted, there is just enough space to fit it in above the ATU toroid in the PFR3B. 3mm LEDs were fitted to the top case and cabling run to avoid pinching at the side of the board. The diode D1 is a BAT43, the LED current limit resistors were changed, R14 was omitted, R15 was 270 , R16 and R17 300 .

Whilst it's not great it does work, I'm sure it can be improved. Thank you to Mike for providing the original inspiration.

[1] http://www.rason.org/Projects/smeter/smeter.htm
Note – we should add that a foundation licence holder cannot make, modify or use a home brew transmitter.

QRP SCRAPBOOK

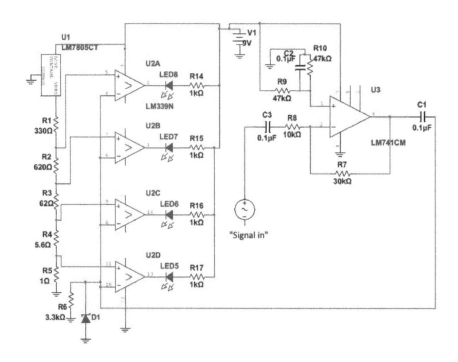

Tidy Toroids
Paul Smith, G4BJG mithsp@gmail.com

I use many inductors, transformers and bifilar/trifilar/quadrifilar windings wound on a T37-2 (and 6), FT37-43 and similar toroids but how could I tidy up the assembly so that they look something better than a spider walking across the PCB?

My latest idea is to mount the wound toroid onto an 8 pin DIL IC socket, preferably of the turned pin variety. I remove any pins not required by drilling them out using a 1.5 or 1.6 m.m. drill. While drilling it helps to hold the body of the individual socket (not the pin) with a pair of snipe nosed pliers, preferably the type with teeth.

Tin the wire ends and leave enough lead length to push them into the sockets and solder. One tip here – push the IC socket into another one. This stops the pins moving at an angle during soldering!

Use a small amount of White/Blutak between the toroid and the IC socket to support it. This method seems to work for many assemblies. A bifilar winding, inductor or tapped coil sits happily on a single-in-line socket strip. I didn't use any Blutak as support on this one but a small amount may help.

I have shown these assemblies made from T37/FT37 and T50 toroids. The T50/FT50 toroids overlap a 0.3" pitch IC socket but I managed to make a reasonable job of it. A T68 toroid might require a cut down 0.6" pitch IC socket.

Switched range constant current charger
Phil Stevens G3SES (philg3ses@gmail.com)

During an attempt to organise the shack I came across numerous Nickel-Cadmium and Nickel-Metal-Hydride batteries. These range from the usual 1.3 V cells to 6V, 9V and 12V units. The rated values were from 280 mA/H to well over 1000 mA/H.

I realised that to check their state I needed a constant-current battery charger which would give the 10 hour charging rate usually used for charging these types of cells and batteries. Not wanting to buy a battery charger I decided that a simple voltage to constant current could be built around the very cheap and available 7805 5V 1A three terminal regulator. The National Semiconductor voltage regulator data book gave me all the information required.

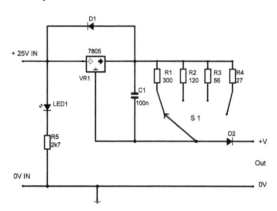

My aim was to have current output ranged in a binary sequence of 25 mA, 50 mA, 100mA and 200 mA with current set by a four-way rotary switch.

The voltage across the 7805 must not exceed 35V and it should be connected to a heatsink with all the terminals insulated from chassis, including that usually called 'ground'.

As I had a suitable mains transformer I built a standard 25V DC supply which would provide at least 0.25A. If you have a variable bench PSU then that could be used.

The output current is determined by the resistor between the 'ground' and 'output' regulator terminals. The formula to determine the value is as follows:

Current Out $I_o = 5V/R + I_q$

R is the value of the resistor in ohms.

I_q is the quiescent current out of the 'ground' and taken as 8 mA

As an example, suppose we need to decide the resistor value for a charging current of 100 mA.

$R = 5V / ((100 - 8) \times 10^{-3}) A = 54$ ohms

The nearest preferred value is 56 ohms. The 27 ohms resistor needs to be 1W rating. The two diodes are for protection of the 7805 and are the usual 1N400X type. The capacitor across the resistor is to ensure oscillation cannot take place.

My prototype was built in a ventilated small metal box and the tab of the 7805 was insulated from the box (heatsink) by a TO220 insulator, screw insulator and silicon grease.

Maximum dissipation occurs when the charger is providing 200 mA into a single cell and is less than 5W in the 7805. If you have any problems or queries please contact me

PART 6 MISCELLANEOUS

VK5TM Noise Canceller
Terry Mowles VK5TM - www.vk5tm.com

The VK5TM Noise Canceller is another version of the design originally developed about 1989 by G4WMX and GW3DIX. Later, Hans DK9NL updated it to include a HF Vox circuit, however, it proved unreliable on SSB transmission and has been removed in my version. A couple of other changes include the use of SMD JFETs and a double-sided pcb.

The operation of the circuit involves the cancellation of local interfering signals, that are picked up on a relatively inefficient 'noise antenna' and are made 180o out of phase with the signal on the main antenna, which also contains the interfering signal plus the wanted signal. After the mixing of the two interfering signals, the main wanted signal remains. Tony G4WIF's blog at http://www.fishpool.org.uk/noisecancel.htm has a nice list of links (supplied by Nick G8INE) to more information about how these work and other variations.

It should be pointed out that this circuit is not a cure for broadband, multiple noise sources. If you are experiencing this sort of interference, you should probably be looking more towards DSP noise cancelling techniques.

There is nothing out of the ordinary as regards the circuit and construction is straight forward but a couple of quick points first.

Without power applied the circuit is in bypass mode (so you won't do any damage transmitting into an unpowered unit).

The power supply input is diode protected to prevent damage from reversing the power connections.

The changeover relays are two 6V 3A contact rated units wired in series. A 220uF capacitor in series with the supply enables quick operation of the relays, while the paralleled 150Ω resistor reduces the hold current once the capacitor has charged.
The bandwidth of the unit is determined by transformer T1, depending on the core used and number of windings. This is an area you can experiment with.
Maximum transmit power through this unit should be 100W or less.
Rather than take up all the space in the magazine with the full testing, adjusting and connecting to your station of the Noise Canceller, you are referred to my website www.vk5tm.com for a more in depth article including parts list etc.

QRP SCRAPBOOK

Schematic

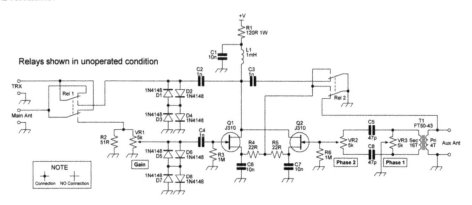

Fig1 - The active part of the Noise Canceller

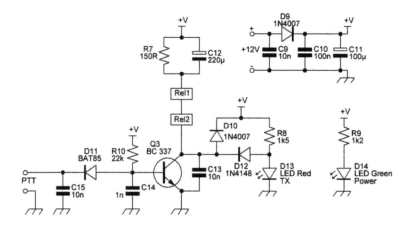

Fig 2 - Power supply and PTT section

Construction

The suggested sequence of construction is to first fit R1, R3 & R6 followed by C1, C6, C7 & L1. Note that R1 & L1 will run warm and should be fitted proud of the pcb by a couple of millimetres to allow airflow around them. These components are fitted first to help prevent static build-up during soldering of Q1 & Q2.

Next fit the two smd devices, Q1 & Q2. After that, fit all the remaining low profile components (resistors & diodes) followed by the remaining transistor and capacitors.

PART 6 MISCELLANEOUS

Relays 1 & 2 can be fitted next followed by transformer T1.

T1 is 16 turns secondary/4 turns primary, 0.2-0.3mm enamelled copper wire, wound on an FT50-43 toroid core (0.5mm wire was used in the prototype and is somewhat easier to handle).

Finally, fit the three potentiometers. An earth wire is then soldered from the earth point on one side of the potentiometers, across the top of the potentiometers to the earth point at the other side. This wire is then soldered to the body of each of the potentiometers.

Hand capacitance has a marked effect on the controls, particularly VR2, as does stray external RF, so this unit should be mounted in a metal enclosure. If you have or can source potentiometers with plastic shafts, they would be better than the metal shaft variety.

Component Overlay

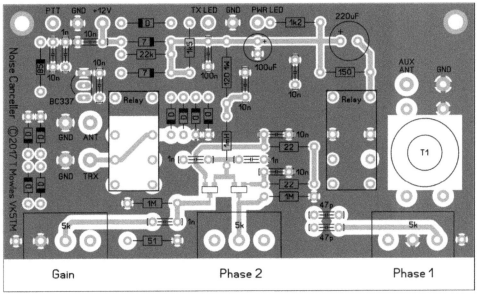

The diodes marked "D" are 1N4148's, "7" - 1N4007's and "85" - BAT85.
(1N4007's were used as that is a standard component in my workshop, 1N4001's, 1N4002's etc should be equally suitable and a 1N5819 was used for the BAT85).
The pcb is available from the author (see website) for $10 AUD (Australian residents need to add GST) plus Post and Packing and a kit of parts, excluding case, connectors and knobs, should be available by the time you receive this edition of Sprat.

Finally, I would like to thank Tony G4WIF, Nick G8INE and Paul VK4APN for their valuable feedback during construction of the prototype version of the Noise Canceller.

QRP SCRAPBOOK

Power supply using a high value capacitor
F6GLZ Jean-Claude Gerwill, G-QRP 7423

For several years, I'm using a more than classic power supply for the workbench. It consists of an LM317 voltage regulator switchable with different voltage outputs. Filtering is done by using recovered parts issued from a power supply (50V / 3A), namely a coil of 1 Henry and an electrolytic capacitor of 18,000 uF as seen on the schematic :

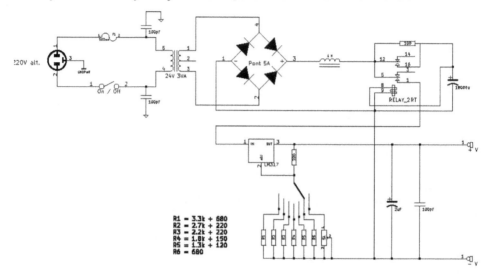

Without the protection done by the relay circuitry, this filter needs a high current when the switch is just going on. This is due to the fact that the capacitor is completely discharged at this moment.

To avoid this high starting current, the capacitor is « slowly » charged during a few seconds through a low value resistor (100 ohms). After a short time, the capacitor is charged enough to excite the relay. The resistor is then shorted, and the voltage is fully applied to the regulator only after this short charging delay.

This protection can be used in any device using capacitors of high values for filtering the output voltage.

G-QRP Membership Application Form

Subscription Rates:-
United Kingdom £6.00, Europe 15.00 Euro (to DX rep) or £12 (to G7ENA)
DX (Including USA) U.S.$22.00 (to USA rep) or £13.00 (to G7ENA)
(All cheques must in Pounds Sterling (GBP), be payable to - "G-QRP CLUB" and must be negotiable on a UK bank.)

The magazine Sprat appears <u>approximately</u> in January, April, July and October. If you join close to those times please expect a delayed response from G7ENA so that she can include the latest Sprat in your welcome pack. The club website www.gqrp.com carries news of Sprat publication dates. We have some special arrangements for members in the following countries: France, Austria, New Zealand, Australia, Belgium, Germany, Italy, Holland, Spain, Denmark and the USA. Otherwise, please pay G7ENA (see below).

Membership Application Form

Tick

I wish to Join the G-QRP Club. I enclose my first subscription. ☐

Membership no.

I wish to renew my membership. I enclose my subscription. ☐ ☐

Callsign _____ Surname and Initials _____

Name used on the air _____

Address – Number and Street _____

Town _____

Post Code/ZIP _____

Country _____

I understand that the club records are maintained on a computer database, and I have no objection to my details appearing on the database.

Signed _____   Date ____ / ____ / ____

Paying By Credit Card

Regretably the club is no longer able to accept direct credit card payments. We can accept online payments using your credit card via "Paypal". Please see www.gqrp.com/paypal/
If you use Paypal you will need to pay in UK pounds £6/£12/£13. The conversion to your currency will appear on your bank statement at the rate effective on the day of the transaction. There will also be a small administration charge.

- We would appreciate an email address so we can contact you if there is a problem.

Email Address _____ @ _____

Send the completed form to:

Membership Secretary:
Daphne Neal G7ENA,
33 Swallow Drive
Louth
LN11 0DN,
England.

Note: Your first subscription Sprat will be the Spring issue of the year that you join. All memberships fall due January 1st. You will receive your renewal notice via SPRAT. UK members are asked to give consideration to changing to a Standing Order so as to lighten the administrative load of the club officers.

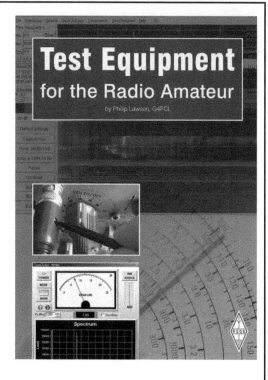

Test Equipment for the Radio Amateur

5th Edition

By Philip Lawson, G4FCL

This book is aimed at the radio amateur, listener and electronic enthusiast who wants to make a variety of measurements without necessarily spending a fortune on expensive test equipment. It is a very practical book, designed to help you develop care and skill in making the most common and important measurements, quickly, safely and affordably.

In this new fifth edition of *Test Equipment for the Radio Amateur*, the reader will find, for the first time, extensive links to internet sources for access to the very latest information on construction projects, equipment and measurements. The sections on commercial and home-brew equipment have been separated for clarity, new items added and some dated items removed. Timeless reference data has been retained; some items of technical theory have been given their own section, and extended, for those who wish to deepen their understanding of these areas. Written in an extremely readable style this book represents a significant update to, and restructuring of, the previous edition.

This book is designed to give an overview of how each item of test equipment works, what it can be used for and even how much it might cost. Many general measurements, plus specific measurements on transmitters and receivers, are described in detail. Matters such as the effect of the test equipment on the circuit to be measured are especially considered, so that the measurement results may be interpreted correctly. A large section of the *Test Equipment for the Radio Amateur* is devoted to home construction, as it is frequently possible to make an extremely useful item of test equipment for a fraction of the price of its commercial counterpart, and there is the added benefit of the practical skill and learning that comes with such a construction project.

Test Equipment for the Radio Amateur is a practical guide to getting the most out of your equipment and understanding exactly how your station is performing. It is simply a must have book for every radio amateur.

Size 174x240mm, 192 pages, ISBN: 9781 9101 9365 5

RRP £14.99

E&OE All prices shown plus p&p

Radio Society of Great Britain www.rsgbshop.org
3 Abbey Court, Priory Business Park, Bedford, MK44 3WH. Tel: 01234 832 700 Fax: 01234 831 496

Index of Articles

4 Bands Dipole (160m, 40m, 20m, 15m) IK0IXI 122
5 in 1 Tester G4UMB 141
5 Watt Linear Amplifier G4DVI 193
6V6 CO/PA Transmitter G3NKS 17
40 Metre WSPR Transmitter G3XIZ 13
40m CW Receiver M0DGQ 51
40m CW Transmitter M0DGQ 5
40m SSB / CW Receiver G3PKW 75
66' Top OCFD for 40 – 6 metres G4LDS 103
455kHz BFO for Older Receivers IK0IXI 178
3560 Transceiver G4UMB 45

A

Add-on Squelch G4UMB 212
Adjustable Loading Coil for Short Loaded Dipole Antennas ... GM4JMU 117

B

Bigger Toy QRP Transceiver .... VK3YE 39, 43
Building the "Sweeperino" G8INE, G4WIF 100

C

Cheap Alignment Tool for Slug-Tuned Coils DJ1WY 218
Cheap and Cheerful Single Lever Paddle WA2JRV 218
Check Your Key Practice DK3WX 223
Chinese X1M QRP Transceiver G0XAR, G5BBM, M0PUB 41
Chirpy 10m Transceiver G3XBM 33
Chopping Board Receiver VK3YE 69
CQ Keyer G4UMB 210
CS5 Direct Conversion Multi-Band SW Receiver G0KJK 96

D

Doublet Experience G4JQT 125

INDEX

E
Easy 4-6 Converter G4KRN 214
Economy Iambic Keyer G4KZO 208
END FED Antenna 80 – 10m L'Olandesina IK0IXI 113
Experimental Loop Antenna for 7, 10 and 14MHz G8ATH/G0SIB 106

F
Foxx with Relay QSK G4UMB 50

G
G4DFV / G4GDR 80m One Valver G4GDR 57
G4GDR 6CH6 Junk Box Special G3VTT 12

H
Hand Generator for Portable QRP IK0IXI 190
Home Made Panel Tip G4COE 11

J
Junk-Box Valve Tester G3VKQ 152

K
KR80 Short Aerial 20 Metre Band QRP Transmitter .... G0KJK 21, 24

L
LATATUN – Spanish Tuna Tin Transmitter EA3WX 8
Low Cost Adjustable PSU G0CVZ 206

M
Magnetic Loops G4KWH 116
Make Your Own Ribbon Cable GM4HTU 160
Modifications to the VK3YE Bigger Toy 40m Transceiver ... G4DFV 43
Modified Limerick Sudden TX for 5262kHz G4GMZ 30
Modular Transmitter G4UMB 19
More on Using the RTL2832U R820T Dongle .... G4WIF 85, 86
Mousetrap Receiver SM7UCZ 64

INDEX

N

N Channel JFET Tester	G4KRN	148
New High Performance Regenerative Receiver	F5LVG	73
No Cost Traps	G4ICP	128
Novel Morse Key	G4UMB	226

O

Observations with Ceramic Resonators in Super VXO Mode	VK6WV	219
Occam's Microcontroller	M0XPD	161
Overvoltage Protection Using MAX6397	G4GIY	215

P

Power Supply Using a High Value Capacitor	F6GLZ	234
Primer for Software Defined Radio (SDR) Using the RTL2832U R820T Dongle	G4IIB	80

Q

Quick and Easy GPS-Locked Frequency Standard	G4OEP	133

R

Real Tuning Knob for SDR Radios	G4WIF	198
Regulated Adjustable HT Power Supply from 12 Volts	G3XIZ	144
Remote antenna Changeover Relay	G4KIH	121
RF 20dB Amplifier for DC Receivers	CM2IR	159
RF Generation for Superhets	M0XPD, N5QW	179
RF Voltage Source Test Generator	VK6WV	143
Rig Safety (antenna static drain)	AF8X	124
RR9 Short-Wave Receiver	G0KJK	66

S

Signal Injector and Tracer	G4UMB	138
Simple 4 Watt Add-on PA for 7MHz Micro Transmitters	G4AQS	199, 200
SIMPLE 40m Transmitter	G4UMB	32
Simple 80m Modular Receiver	G4UMB	92
Simple DC RX Covers 8 Bands 40m to 10m with 27 Components	G6JLH	93

INDEX

Simple DDS VFO	VK5TM	188
Simple Electrolytic Capacitor Tester	G4UMB	150
Simple QRP 4m Transverter	G3XBM	156
Simple Smart Continuity Tester	G4UMB	149
Simple Upside Down Magnetic Loop	G4IIB	111
Simple, Cheap S-Meter for Portable rigs	M6SLO	227
SLA Charger	G3XGY	221
Small Talk 160m Transmitter on FM	G4UMB	29
Sudden PSK?	G0FUW	62
Switchable Crystal Oscillator	M0XPD	201
Switched Range Constant Current Charger	G3SES	230

T

That RTL2832U R820T Dongle Again	G4WIF	86
Three "Moxo" (Modified OXO) QRP Transmitters	G0KJK	25
Through a Double Glazed Window	G4FYY	187
Tidy Toroids	G4BJG	229
Tin Can 2 (Two Transistor 80m Receiver)	VK2ASU	89
'Tiny Toy' – a 40m CW QRP Transceiver	VK3YE	35
Trap Dipole Antenna for 5262kHz and Beyond	G4GMZ	130
"Twelvevolter" – A Hybrid Receiver for 40m	G4DFV	58
Two Band Ceramic VXO	VK3YE	191

U

Ultimate Bipolar VFO	G4FQN	154
Universal SSB Generator	G4COE	171
Using Ceramic Resonators in SSB Filters	G4HJW	211

V

Valve Regenerative Receiver	M0DGQ	77
VFO and signal Generators Using the Si5351A Chip	G0UPL	139
VK5TM Noise Canceller	VK5TM	231

W

Wireless Side Tone	G0VXG	207
WSPR Frequency Calibration Tool	G8FDJ	136